BRA

TRAINING

How to Shape Your Plastic Brain by Forming New
Connections

(Improve Memory, and Get Smart Using Brain
Plasticity)

Michael Brown

Published by Tomas Edwards

Brain Training: How to Shape Your Plastic Brain by Forming New Connections (Improve Memory, and Get Smart Using Brain Plasticity)

ISBN 978-1-990268-23-6

Legal & Disclaimer

The information contained in this book is not designed to replace or take the place of any form of medicine or professional medical advice. The information in this book has been provided for educational and entertainment purposes only.

The information contained in this book has been compiled from sources deemed reliable, and it is accurate to the best of the Author's knowledge; however, the Author cannot guarantee its accuracy and validity and cannot be held liable for any errors or omissions. Changes are periodically made to this book. You must consult your doctor or get professional medical advice before using any of the suggested remedies, techniques, or information in this book.

Table of Contents

Introduction

Congratulations on downloading this book and thank you for doing so.

The following chapters will discuss what it means to overthink and what you can do to change it. You will learn about the symptoms of overthinking and how it affects your day-to-day life, before going on a step-by-step journey to alleviate the stress of overthinking and negative thought processes.

Many people overanalyze and overthink every aspect of their lives, whether it's their jobs, their relationships, lack of fulfillment, or a constant stream of stress. All of these struggles seem inevitable in a normal life, and if you look around, people seem to have accepted that life and stress and negativity will always be connected, even in the best-case scenarios.

Forming good habits like mindfulness meditation, positive relationships, and adequate sleep will lead you to a place where the negative habits that lead to overthinking fall away. From poisonous relationships to a cluttered living space, shedding the things that hold you back will give way to a whole new you, ready to meet life's challenges with a mind filled with positive thoughts and meaningful goals. It is my hope that by working every day to form better habits, you will come to realize your full potential without feeling like overthinking is an inevitable part of life. You can take back control!

There are plenty of books on this subject on the market, thanks again for choosing this one! Every effort was made to ensure it is full of as much useful information as possible, please enjoy!

Chapter 1: What Your Brain Needs To

Know About Neuroplasticity

Your life truly shapes your cerebrum. Your cerebrum changes both structure and capacity consistently in light of your life encounters, feelings, practices, and even musings. This stunning transforming capacity, called neuroplasticity, is both good and bad. Neuroplasticity has enabled individuals to make momentous recuperations and beat genuine psychological wellness challenges. It's the manner by which all learning and memory occur. Neuroplasticity has sweeping ramifications and potential outcomes for pretty much every part of human life and culture.

Notwithstanding, neuroplasticity is additionally the explanation that some emotional well-being and cerebrum issues appear in any case. It's similarly as simple to corrupt your mind as improve it. This is what your mind has to think about neuroplasticity.

What Is Neuroplasticity?

During the 1980s, scientists at The University of California at San Francisco (UCSF) affirmed that the human cerebrum rebuilds itself following the "Hebbian rule." Donald Hebb, a Canadian clinician, first suggested that "Neurons that fire together, wire together" implying that the mind ceaselessly adjusts itself physically and operationally dependent on approaching improvements.

Neuroplasticity is really an umbrella term alluding to the numerous abilities of your cerebrum to rearrange itself for the duration of your life because of your condition, conduct, and inner encounters. Science used to accept that the mind was just variable during specific periods in adolescence. While the facts demonstrate that your mind is significantly more plastic in the early years and limit decreases with age, versatility occurs for the duration of your life from birth until death.

Science has affirmed that you can get to neuroplasticity for positive change in your own life from multiple points of view at any age. Saddling neuroplasticity in adulthood isn't exactly as straightforward as a portion of the neuro-promotion would have you accept, yet it can most unquestionably be practiced under specific conditions.

How Neuroplasticity Helps Your Brain

Neuroplasticity has empowered individuals to recuperate from stroke, damage, and birth anomalies, improve indications of mental imbalance, ADD and ADHD, learning handicaps and other cerebrum deficiencies, recoup from sadness, tension, and addictions, switch fanatical impulsive examples, and that's only the tip of the iceberg. Because of neuroplasticity, you are not stayed with the mind you're brought into the world with or that you have at some random time in your life.

Neuroplasticity has conceivable positive applications in numerous territories, including medication, psychiatry, brain research, connections, instruction, and then some. Where it stands to have the most potential is for the person in their own life. Because you can figure out how to deliberately control your reasoning, responses, and conduct, and a portion of the encounters you have, you can regulate your own "self-coordinated neuroplasticity" and welcome change and mending into your very own life.

You can reconstruct a messed up mind

I utilized neuroplasticity to reconstruct my mind after genuine damage, the aftereffect of a suicide endeavor. Throughout the years, I picked up all that I could about neuroplasticity and got down to business revamping my cerebrum through exercise, psychological wellness instruments, contemplation, representation, and care rehearses. Through the wonder of neuroplasticity, I relearned how to talk, compose, move with coordination, recall, and think obviously once more. (Peruse my story here.) This time around I purposefully shaped a mind that was increasingly positive, versatile, quiet, and glad.

Because each mind damage is different, the conceivable outcomes for recuperation are as well. Be that as it may, deliberately tackling neuroplasticity can help restore the cerebrum after damage.

Numerous Mental Health Conditions Are Reversible

It is conceivable to beat an emotional well-being condition, for instance, a fear, discouragement, schizophrenia, by driving a cerebrum back towards ordinary activity through neuroplastic change. In his book, Soft-Wired: How the New Science of Brain Plasticity Can Change Your Life, Dr. Michael Merzenich, one of the establishing scientists at UCFS, clarifies:

… we realize that treating these clutters with medications is basically not good enough. A broken mind is certifiably not a synthetic stew that is feeling the loss of a key flavor. Cerebrum science is incredibly muddled… .The thought that a solitary medication can give the sole premise of treatment for all that messed up apparatus in a mind boggling condition like schizophrenia is evidently ludicrous. There is a far more noteworthy prospect for accomplishing neurological recuperation through mind preparing intended to switch many misshaping changes that are a piece of the disease – and through that preparation renormalizing the cerebrum physically and artificially."

Concentrates on cerebrum pliancy directed by Merzenich and his associates and different researchers around the globe have on the whole exhibited that numerous parts of your intellectual competence, knowledge, or control – in typical and neurologically disabled people – can be improved by serious and suitably focused on social preparing.

Neuroplasticity can help or damage you. It's your decision.

Neuroplasticity permits learning and memory

Learning and memory rely on neuroplasticity in the associations of mind circuits. Both are neuroplastic forms, which means they include concoction and auxiliary changes in your mind.

By adjusting the number or quality of associations between synapses, data gets composed into memory. It used to be accepted that a solitary memory is comprised of slight shifts in specific neural connections in a specific arrangement involving that memory. Nonetheless, ongoing examinations show a lot more prominent multifaceted nature than just changes happening in single neural connections. Or maybe, changes seem to happen in enormous disseminated arranges all through the cerebrum.

Reviewing a memory is likewise a neuroplastic procedure. Each time you review one, your cerebrum reconsolidates this procedure fusing and separating it through what your identity is, the thing that you know, and your outlook at the hour of recalling. Consequently, memory is a functioning and continuous procedure, and as per Jonah Lehrer, in his book, Proust Was a Neuroscientist, "A memory is just as genuine as the last time you recollected that it."

How Neuroplasticity Hurts Your Brain

When you find out about neuroplasticity, it's for the most part related to wonderful, positive mind and life changes as itemized previously. Nonetheless, this equivalent trademark, which makes your mind incredibly strong, likewise makes it truly defenseless against outside and inward impacts.

This is because of neuroplasticity that bad habits become imbued in your mind, significant aptitudes are lost as your cerebrum decays with age, and some significant cerebrum ailments and conditions appear in humans. For instance, sadness is essentially a mind example scratched into an individual's cerebrum after some time through neuroplastic changes.

The Plastic Paradox

In his book The Brain That Changes Itself: Stories of Personal Triumph from the Frontiers of Brain Science, Norman Doidge considers pessimistic neuroplasticity the "plastic oddity." He composes:

The plastic Catch 22 is that the equivalent neuroplastic properties that enable us to create progressively adaptable practices can likewise enable us to deliver increasingly unbending ones. All individuals begin with plastic potential. A few of us form into progressively adaptable kids and remain as such through our grown-up lives. For others od us, the suddenness, innovativeness, and flightiness of youth offers path to a routinized presence that rehashes a similar conduct and transforms us into unbending personifications of ourselves. Anything that includes unvaried reiteration — our professions, social exercises, aptitudes, and mental issues — can prompt unbending nature. To be sure, it is because we have a neuroplastic mind that we can build up these inflexible practices in any case."

Two Primary Ways You Can Drive Neuroplasticity

Because neuroplasticity adheres to the Hebbian rule, it's in a general sense reversible. Neurons that fire together wire together, yet neurons that don't, won't. You have a "use it or lose it" cerebrum. Data once in a while got to and practices only from time to time rehearsed cause neural pathways to debilitate until associations might be totally lost in a procedure called "synaptic pruning." Neuroplastic change happens in light of upgrades prepared in the mind which can start either inside or remotely.

Remotely Driven Change

From youth through adulthood, the occasions of your life shape your cerebrum. As meager individuals develop, associate with others, and investigate the world, associations are wired in their cerebrums dependent on their encounters. When you're youthful, the vast majority of what happens is out of your control. As grown-ups, our cerebrums are impressions of our day by day schedules. Your habits, both good and bad, actually get wired into your cerebrum. Approaches to drive remotely determined change would be:

Take a stab at something new. This can be as included as learning a different language, returning to class, taking move classes, or acing a melodic instrument. It additionally can be as straightforward as evaluating another café, perusing a book out of your ordinary type (particularly fiction), or tuning in to a new style of music.

Blend things up. Use your nondominant leg to fire up the stairs or your nondominant hand to eat or brush your teeth. Move the mouse to the opposite side of the console. Rest on a different side of the bed. Take another course to work. Get your cerebrum off of programmed.

Mood killer the GPS. Use a guide and your mind. You could even get purposefully lost and attempt to discover your way back without utilizing your GPS or a guide. (There's an entire game like this called orienteering.)

Exercise in another manner. Attempt an entirely different movement. If bicycling, shift your courses and territories. If running or strolling, get outside when conceivable, overlook the earphones, and change surfaces, ways, and landscape. The thought is to get your mind in the exercise.

Train your cerebrum. I know there is an entire discussion about whether cerebrum preparing works or not. There is no doubt as far as I can say or Dr. Merzenich's. It helped me recuperate from my mind damage. Dr. Merzenich proceeded to begin an organization, Posit Science, which offers cerebrum preparing dependent on neuroplasticity and science (Brain HQ). Not all cerebrum preparing is the equivalent.

Travel. Travel to another city, another nation or right not far off. A switch of landscape awakens your mind, sparkles inventiveness, and can even lift joy. Another condition difficulties your mind and takes it off auto-pilot. You need to consider even little things when in a new spot.

Be social. Converse with individuals eye to eye. Take part in discussions and truly tune in. Make new companions dissimilar to any you as of now have. Higher social commitment is related with higher psychological working and decreased dangers of subjective decay.

Inside Driven Change

To energize neuroplastic change inside intends to impact the tasks of your cerebrum through working with your brain. As per Merzenich in the video Make Your Brain Smarter Every Day @ Any Age, mental exercise drives pliancy the same amount of as outside movement does.

Your psyche shapes your cerebrum. All that you think, expectation, feel, and envision physically changes your mind – regardless. You can deliberately outfit this procedure for your advantage. Daniel G. So be it, MD writes in You Are Not Your Brain: The 4-Step Solution for Changing Bad Habits, Ending Unhealthy Thinking, and Taking Control of Your Life:

… we can really use the psyche to change the cerebrum. The straightforward truth is that how we concentrate, how we purposefully direct the progression of vitality and data through our neural circuits, can legitimately modify the cerebrum's action and its structure."

Approaches to Drive Neuroplastic Change Internally

1. Care

The exploration about the positive effect of care on the cerebrum and emotional wellness focuses to neuroplasticity as the cause. In care, by purposefully coordinating consideration internal and developing attention to the breath or contemplations and emotions, you are getting to be mindful of your cerebrum's Default Mode Network (DFM) and applying power over it. When you intentionally guide your DFM, you're interfering with habitual idea designs and situating your mind right now.

2. Reflection

Alongside the numerous scientifically demonstrated advantages of reflection for your mind, it expands neuroplasticity. Reflection has been demonstrated to diminish pressure, nervousness, and misery, which have been appeared to constrain neurogenesis, the introduction of new synapses. What's more, you don't have to contemplate for a considerable length of time to begin receiving rewards either. One examination demonstrated cerebrum changes after only two months of ordinary reflection.

3. Perception

Neurons fire and synthetic substances are discharged in your mind in the case of something is genuine or envisioned. On cerebrum checks, innovative contemplations actuate numerous indistinguishable mind zones, which legitimately impact you, physically and inwardly. From a neuroscientific point of view, envisioning a demonstration and doing it are not unreasonably different. Thus, perception enables you to give your creative mind something to do for you to change your cerebrum. Research has approved that the training impacts physical changes from muscle solidarity to mind pathways.

Chapter 2: Reason Behind Short Term

Memory Loss

Memory is considered to be a very complicated process since there are all kinds of ways because of which a human memory may work well or not. Memory is considered to be the primary factor behind everything we do such as:

Remembering somebody's name or to remember a phone number

From remembering information that you need to clear an exam

To remember how to walk or how to speak

Memory makes up our repeated experience of life and provides us with a sense of self-awareness. Hippocampus plays a vital role in the memory and since both sides of the brain are symmetrical, we can find the hippocampus in both the hemispheres. If either side of the hippocampus is destroyed or damaged, as long as the other side is undamaged, memory function will remain nearly normal. If both sides of the hippocampus are damaged, it could obstruct the ability to structure new memories, called "anterograde amnesia".

With your age (as you grow old), functioning of the hippocampus could also drop down. People may have lost as much as 20 percent of the nerve connections in the hippocampus by the time they reach their 80's. Fortunately, this neuron loss is not applicable for seniors.

Experts accept as true that you could hold about seven items in short-term memory approximately for 20 to 30 seconds. However, the majority of the memories (short-term) are forgotten quickly and the ability of storing the short-term memories is quite limited. By using memory approaches such as "chunking" this capacity can be stretched to some extent. It involves combination related information into slighter "chunks".

For storing a list of items, the capacity of short-term memory was somewhere between five and nine. But at the moment, a lot of memory experts agree to the fact that the accurate capacity of short-term memory is more likely nearer to the number four (on a scale of 1 to 10).

You would be able to see this in action for yourself just by trying out this short-term memory experiment. Spend at least two minutes of time and try to remember a random list of words and then take a blank piece of a paper and try writing down as many words as you can.

Testing yourself on the information really helps you to remember it in a better way. Here is the list of some major reasons why we are not able to remember the information and forget it easily:

Retrieval Failure: One likely explanation of retrieval failure is also known as "the decay theory". As per this theory, a memory trace is created each time a new theory is formed. This memory trace starts disappearing and fades as the time passes. If the information is not rehearsed and retrieved, it would be lost.

However, the only problem with this theory is that if the memory have not been remembered or rehearsed are amazingly stable in long-term memory.

Interference: This theory suggests that some of the short-term memories actually interfere and compete with other memories that are stored in the human brain. Interference is more likely to occur only when the information that was previously stored in memory is very similar to other information. There searchers have identified two types of interferences, these are:

Retroactive interference: normally occurs when new information gets in the way with your ability to remember previous learned information.

Proactive interference: normally occurs when an old memory makes it more impossible or difficult to remember a new memory.

Failure to store the information: There are times when losing information has more to do with the fact that it never made it into long-term memory in the first place and less to do with forgetting. Encoding failures sometimes may stop the information from going inside the long-term memory.

Motivated Forgetting: Occasionally, we might vigorously work to forget the memories, particularly those of traumatic, disturbing events or experiences. The two basic types of motivated forgetting are: repression (an unconscious form of forgetting) and suppression (a conscious form of forgetting).

"How erratic and slow is the growth of a student who cannot even keep in mind what he has learnt".

On the other hand, people with a good working memory are considered to be more self-assured and optimistic, and more likely to direct a successful and happy life.Hence, the use of mnemonic devices could improve the memory a lot, especially the recall of long lists of numbers, names etc.

Chapter 3: Overhaul Your Attitude And

Shape A More Positive Mind

Having read the previous chapter, you now know that thanks to the pliable nature of your brain, you have the potential to learn new skills. In Chapter 3, we will take this idea further and look at some practical steps you can take to help you maximize your intellectual ability. However, neuroplasticity isn't just useful in improving your vocabulary or learning to play a musical instrument. It's also your passport to the development of new life skills and a whole new outlook on life in general. In this chapter, you will come to understand how your brain's malleability is an enormous asset when it comes to making changes in your approach to life.

Positive thinking is a skill

You may believe that some people are simply born positive thinkers, and that this element of a person's character is largely fixed. Whilst it is true that there is a heritable component to personality traits, there is increasing evidence that if you are willing to put in the effort to change your thinking patterns, both your brain and behaviors will help move you in a more positive direction. People who have mastered the art of positive thinking are more likely to be successful in all areas of their life. Note that being a positive thinker does not mean that you wilfully discount constructive criticism or refuse to face up to reality. Instead, it simply means that you are willing to believe in yourself and the possibility that most situations contain at least the potential seeds of a good outcome. Just as driving around London repeatedly helps develop neural networks in taxi drivers responsible for successful spatial navigation, repeated practice at positive thinking helps you automatically search for the upsides whenever you hit a roadblock in life. This ultimately helps you build resilience, which is protective against depression and helps you overcome the challenges life throws at you.

Depression, neuroplasticity, and the practice of positive thought The effects of depression on the brain have been invaluable in developing our understanding of neuroplasticity and how it relates to mood and cognitive function. Recent research has demonstrated that depression is not just a mental state or experience, but a neurological event that has far-reaching consequences beyond a feeling of being "down" or "sad." Depression is a mental illness characterised by a number of psychological, emotional and physical symptoms including feelings of sadness, feelings of hopelessness, thoughts of death or suicide, loss of interest in previously-enjoyed activities, changes in weight and appetite, vague or random aches and pains with no discernible physical cause, and a tendency to withdraw from social situations. Psychologists believe that there are several underlying factors that may cause or at least contribute to depression, including genetic predisposition and neurochemical imbalance. Some of these may be beyond your conscious control. On the other hand, there is also plenty of evidence to suggest that a significant factor that keeps depression going is an individual's thinking style. Put simply, people who suffer from depression tend to see the world in a maladaptive way that keeps them locked in a cycle of negative thinking, negative actions, and withdrawal from the world around them. To break the feelings of bleak pessimism that often accompany depression and keep it going, it's important that a depressed person retrains their brain to interpret external events in a more constructive way. This is the rationale behind a type of psychotherapy known as Cognitive Behavioral Therapy, or CBT. The principle of CBT is as follows: It isn't just what happens to us in life that can make us feel a particular way – what ultimately determines how we feel is the meaning we personally ascribe to external events. CBT practitioners believe that a key difference separating those with depression from those who are mentally healthy is that the former group habitually fall back on negative ways of viewing the world. In other words, when you are depressed, you continually teach yourself that the world is a bad place, that you will get hurt on a regular basis, and that there is little reason to think that things will get better. This style of thinking becomes "normal," and over time you may not even realize just how deeply your pessimistic your thinking style has become entrenched. By now, you know that neurons firing together tend to wire together. Think negative thoughts on a regular basis and they will come to represent your reality. This has discernible effects on the brain. Studies comparing the brains of people with and without depression have found that mental illness can induce a state of "negative neuroplasticity" in which certain thinking and behavioral patterns become entrenched and maintain the symptoms. Research carried out at the University of Michigan using a brain scanning method known as positron emission tomography

(PET) found that people with untreated depression have significantly fewer serotonin receptors than those not diagnosed with the condition. This is important because in order to feel happy and regulate our moods, our brains need to be able to make proper use of this neurotransmitter.

Other findings have revealed that depressed people tend to experience shrinkage of the hippocampus, which in turn leaves them vulnerable to problems in mood regulation and reduced memory function. The more episodes of depression an individual suffers, the greater the extent of the hippocampal damage. Given that depression is a risk factor for the development of Alzheimer's disease, and that the hippocampus is among the first parts of the brain to be damaged in patients with this condition, it seems likely that the key to this link is an impaired hippocampus. More research is needed in this area, the takeaway message here is simply that depression changes the human brain.

Fortunately, there is also plenty of evidence that psychotherapy, in which people suffering from depression are taught new ways of seeing the world and to challenge their negative thoughts, is an effective treatment. Even if you have been depressed or prone to negative thinking for many years, there is reason for hope – our brains never lose their plasticity, so with the right intervention and behavioral change, you can reverse the damage. Exercise: Challenging Negative Thoughts Whilst negative thought patterns are especially common in depression, most of us feel their effects from time to time. If you allow them too much mental airtime, however, you can begin to feel your attitude towards others and life in general become less optimistic. This can have a destructive effect on your everyday behavior. This exercise is often used by CBT therapists.

1. Identify a negative thought that you find yourself thinking on a frequent basis. This may relate to your social situation (e.g. "I have no friends"), your self-image ("I'm so

incompetent and can't do anything") or about the world in general ("Everyone is so selfish and only out for what they can get"). Write it down and rate its believability on a scale of 1-10, with 10 indicating that you accept this thought as being absolutely true.

2. Now it's time to play detective. What evidence do you have that this thought is actually true? Write down your evidence in favor of this thought. Now look at it from another angle – if you were to present the other side of the case, what evidence could you put forward in support of the motion that this perspective simply isn't true? For example, if you wrote down "I'm so incompetent and can't do anything" in Step 1, it's time to acknowledge that whilst you may not be as proficient in a certain field as you would like, you have succeeded in various other areas and have mastered other skills.

3. Once you have reviewed the evidence, re-evaluate how far you accept your negative thought as being an accurate reflection of reality. If you have taken the

time to generate evidence for both sides (i.e. the statement being true and untrue), you should find that its power over you has been lessened.
4. As an additional step, think about whether or not your negative thought is actually helpful. Even supposing it is completely true (which is unlikely), what do you gain from holding onto it as though it were valid? Destructive thought patterns seldom inspire positive change. Consider the benefits of giving yourself permission to think differently. This exercise is effective because it forces you to realize that the world is not black and white, and that even your most cherished negative beliefs are not immune to the power of critical thinking! If you repeat this exercise whenever you feel yourself slipping into negative thoughts, you will soon train yourself to challenge negativity rather than tolerating it or accepting it as your "normal state." From now on, promise yourself that you are not going to sit back and accept the unnecessary negativity that your brain

throws at you. Instead, pledge to see unhelpful negative thoughts as bad mental habits that you can correct with patience and   effort. Some people find that keeping a journal helps them identify and tackle negative thinking. It isn't realistic to try and write down every negative thought that crosses your mind, but you can certainly make use of your journal in noting down particular themes and devising alternatives to recurrent negative thoughts. Journaling is much easier to stick to when you make it a habit. Remember, when you repeat an action many times, your brain will come to expect it and you will feel uncomfortable if you deviate from your routine. You probably brush your teeth every morning and evening without having to think about it. That's because you have repeated the same action hundreds of times before. The same can become true of journaling or any other new habit that you want to instil. Set a time and place for your journaling practice and commit to it for 30 days. We'll come back to goal-setting and habits

Why "Fake it 'til you make it" really works – the Facial Feedback Hypothesis
The old cliché of "Fake it 'til you make it!" may sound trite and unhelpful if you feel down, but psychologists have discovered that this phrase has merit. Most of us think that our emotions and body language are related as follows: We experience an event or memory, a particular emotion is triggered within us, and our body language changes as a result. We smile when we are happy, sit slumped in our chairs when we are feeling despondent, and so on. Whilst this is true, did you know that there is plenty of research showing that it also works the other way around? You smile because you are happy, but if you make a conscious decision to smile despite feeling grumpy or low, you are likely to feel a genuine uptick in your mood. This phenomenon, whereby the brain receives feedback from facial expressions and other forms of body language in such a way that

the individual actually feels their mood or attitude shift, is known as the Facial Feedback Hypothesis. Early studies in this area used volunteers who were asked to hold a pencil between their teeth as they looked at cartoon strips. Those who were asked to hold a pencil horizontally between their teeth rated the cartoons as being funnier than the volunteers who were asked to hold a pencil in such a way that did not force their lips into a smile. The researchers concluded that because the horizontal-pencil group were in effect made to adopt a "happy" expression, their brains interpreted the cartoons as especially amusing. Other studies have shown that adopting "power poses" (such as standing with your legs apart, back straight and with your hands on your hips) is a genuine confidence-booster in nerve-wracking situations.

These findings are exciting because they imply that we already have the neural pathways in place that allow our brains to associate certain gestures, facial expressions and posture with particular mood states. Why not take advantage of these inbuilt circuits?

Exercise: Try the Facial Feedback Hypothesis For Yourself

The more often you encourage yourself to adopt positive body language, the more natural it will feel, and the more you will strengthen the link between physical movement and your mental state. Getting into the habit of sitting up straight and smiling, even when you don't feel like it, will mean that you have ready access to a quick mood boost wherever you are. So the next time you need a pick-me-up, put the Facial Feedback Hypothesis into action.

Mindfulness and its effects on the brain

Over the past decade, "mindfulness" has become a common buzzword in psychology and psychiatry circles. However, it has proven itself to be more than a passing fad. It turns out that practicing mindfulness results in changes to the brain that can help you feel less stressed, maintain a healthy perspective on events in your life, and reduce your susceptibility to depression. Mindfulness also enables you to make better decisions, which holds exciting implications for those working in a number of fields including crime prevention, education, and social work.

What exactly does "mindfulness" mean? At its most basic, mindfulness is simply the act of focusing your attention on the present moment. It means being willing to stop worrying about the future or ruminate about past events. When you behave mindfully, you are living with full awareness of where and who you are right now. You can practice mindfulness anywhere.

Exercise: Mindful Washing Up

When you next need to wash a dish or spoon, take the opportunity to do a mindfulness exercise. As you run a bowl of soapy water, pay attention to the feel of the water on your skin, the scent of the washing-up liquid, and the sound it makes as it swirls around in the bowl. Feel the texture of the object you are washing against your fingertips. If you feel your attention wandering, notice that your focus has drifted before bringing it back gently but firmly to the present. Mindfulness is often linked with meditation, a broad term that refers to the act of channeling your attention on a particular concept, object, or just your own breathing. Neither mindfulness nor meditation are exclusively the preserve of the religious. Although meditation is popularly linked with Buddhism and Hinduism, people of all faiths and none are able to benefit from practices that allow them mental breathing space. How does neuroplasticity come into this equation? Studies with people who meditate regularly show that meditation literally shapes the human brain in such a way that enhances mood and wellbeing. For instance, a study published in the journal **Neuroreport** explained that compared with those who never meditated, people who practiced regularly tended to have thicker cortices. This was especially apparent in the areas of the cerebral cortex responsible for sensory processing and focused attention. (If you want to retain more of what you learn and focus for longer at school or work, meditation is a great way to improve your performance.)

Not only that, but long-term meditation also helps you exert more control over your emotions. This doesn't mean that regular meditators turn into robots, just that meditation helps equip you with the power to keep strong emotions in check and make more conscious, controlled choices when it comes to responding to challenging situations.

Exercise: Mediation,   Part   I

Simple breathing meditation is a popular exercise for beginners. All you need to do is sit or stand in a comfortable position. Close your eyes and focus on your breathing. Count your breaths slowly. Keep all your attention focused on this once simple task. If your mind starts to drift, bring it back to your breath.

Exercise: Meditation,   Part   II

If you find it hard to sit still for any length of time or dislike focusing on your breath, walking meditation could be a great alternative. Pick a quiet room or area outside, and simply walk up and down in a straight line. Keep your walking pace even. Focus your attention on the sensation you can feel in your feet when they make contact with the ground. When you get into the habit of challenging your negative thoughts and meditating every day, you will soon find that your brain responds favorably to your new habits. Although you might not be able to see the changes on an MRI or PET scanner, you can be sure that your brain's malleable nature will result in significant changes in your mood and overall outlook. Just think how much calmer and more pleasant life will be once you learn to exert greater control over your moods and emotional responses! Start by meditating for 2-3 minutes per day, then gradually build up to 20-30 minutes each morning or evening.

Chapter 4: Knowing Who You Are And

Owning It

Who are you? Really, who are you? Are you your job? Are you a parent? Are you a spouse? Are you successful? Are you a liar? Are you a murderer? Who are you? Personally, I think I could be any of the above if I wanted to. So, back to the question at hand, who are you? Or, what or who makes us who or what we consider us to be?

Do you contend that it is your environment that makes you who you are.Maybe we decide that our current environment would foster a certain "us" that we wouldn't otherwise envision. Perhaps we are just reacting to stimuli-life affects us, we respond, we have no choice. In fact, I would say it is contrary, that we make a choice to not have a choice.

First, we must decide that we are at cause in our lives, not effect. We cause change, we cause the mess we are in, we cause the way we clean it up, we cause the success and we cause the rewards that we reap. But I am getting away from the original question- "Who are you?" We are a lot of things to a lot of different people, corporations, governments, and scientists. We are also a lot of things to ourselves. It is this area that will create the most change when we decide who we are.

Are you worthy? Are you successful? Are you beautiful? Are you smart? Even then, just because you believe that you are who you are doesn't necessarily make it true. You came up

with "I am not worth it" at some point in the past that you do not even remember consciously, that most likely had nothing to do with you in the first place, if it did, you most likely did not have the tools to make the right decision about who you were and now are.

Are we more than our thoughts? Well, yes. But that is where it starts. Walk around any department store these days or check out social sites and you will be inundated with sayings about thinking and believing and becoming and thoughts and dreams, etc. In NLP, the same holds true, your thoughts are a result of your values, your values determine your beliefs and your beliefs drive your outcomes. The nice thing about all of this is that the past is the past. It is just that- it has passed.

You are no longer the person that you were when you were five or ten or 15 or 20 or 30 or 50. Times change, you change, your thoughts change. You are who you are now!

now...now.... now, and now..... now, get it? And who you choose to be now will determine the path that you will begin to go down to create your future. The possibilities are endless. There are endless "you" waiting to be created, you get to pick.

You can reinvent yourself at the start of every day, at the end of every lunch, before every dinner, and right before you go to bed! Your life is in YOUR hands, and your hands alone. You have the power to be whoever you want to be, but you have to focus to do it. If you want to be happy don't look outside, look within. Start by smiling. Then look in a

mirror and smile. Then look at someone and smile, they WILL smile back or at least be a little confused.

If you want to be happy, focus on being happy. Focus on what you want in life not what you DON'T want. If you are tired of being fat, don't focus on all the reasons why you are fat, rather, focus on all the ways you can be healthy and fit.

Then do those things that will make you healthy and fit.

Remember when you were little, you didn't settle for being who you thought you were, you were anything that you could imagine.

It's time to imagine again, it's time to create that amazing person lurking beneath the surface of who you currently believe that you are. Decided who you want to be today and be it.

Chapter 5: How Do Habits Form?

Now that we understand a bit more about habits and how they work together with the routines that we set, it is time to take a look at the process that happens when habits start to form. No matter what kind of habit we are forming, the brain is going to treat them all similarly. Understanding how this process forms, and why it is so strong, can help to give us a deeper understanding of our habits.

Remember that habits are formed because they make things easier. The brain would get exhausted pretty quickly if it had to focus on every mundane task that you had to do during the day. Without habit formation, the brain would have to be actively involved, even in tasks that you do daily. It is much easier if the brain is able to take many of those tasks and automate them. So any task that you do, the brain is already going to actively look into whether it can become a habit, and whether or not you can move it from the cerebral cortex over to the basal ganglia to make things easier.

Several processes have to occur before an action, or a series of actions becomes a habit. We will take a look at some of these processes below, so we better understand how habits are formed.

The neurological loop

Now we need to take a look at how habits are going to form. To do this, we need to understand the neurological loop, which is sometimes known as the habit loop. This loop was discovered by researchers from the Massachusetts Institute of Technology or MIT while they were analyzing how rats were able to run through mazes.

After this research, they discovered that there was quite a bit of cerebral cortex activity during the first few runs that the rats did through the maze. But the more that the rats went through this maze, the less activity that arrived in the cerebral cortex. But this didn't necessarily mean that the memory was the culprit of this. In fact, the researchers found that as the rats went through the maze more times, there

45

was also a decrease in how much brain activity occurred in the areas that governed memory.

With further researcher, they found that the more the brain is exposed to an action or pattern that is repeating, the brain will move the activity over to the basal ganglia instead so that the cerebral cortex can focus more on higher and more intensive functions. This part of the brain, the basal ganglia, is going to be mostly responsible for motor control along with having a roll in emotions, executive behaviors and functions, and motor learning.

Due to this study, the researchers at MIT were able to simplify this neurological loop and assigned it as the core of every habit that we develop. Three main parts come with this neurological loop including a cue or a trigger, a routine, and a reward:

Cue: This is going to be the trigger or the thing that causes the routine, or will kick in the automatic urge.

Routine: This is the behavior that you would like to change.

Reward: Craving that is needed.

Often the habits that we form are going to be automatic. We might take some time to work on some habits, like developing a habit of brushing our teeth, but there are some habits that we just fall into automatically because they make life easier. This is part of why breaking a bad habit can be difficult. We have started to do a habit automatically, without even thinking about it, so taking the time to actually think about the actions and then working to fight against the automatic actions of the body, can be hard.

There are four steps that anyone can use in order to control this neurological loop. These steps will include identifying your routine, experimenting with the rewards that you receive, isolating the cue, and coming up with a plan of action. Let's take a look at each of these parts and how they can be used to help you take control of the neurological loop that occurs in your brain.

Identify the routine

When a certain cue occurs, the routine is going to be the way that you react automatically to it. For example, if you find that every time you see someone else with a soda, you get up and go and get yourself one, then this is your routine. There is usually some kind of cue or trigger that makes the reaction possible, without us thinking about it.

When you want to control your neurological loop, you need to know what the cue or the triggers are. While you may find that it isn't the most realistic thing to try and change each and every habit that you have now (remember the brain has worked to form a ton of habits to make things easier), if you start acknowledging these cues and triggers now, you are going to be more prepared to change a habit when you are ready to get with it.

Experiment with the rewards

This section is going to need a little time and some guessing to find out what is going to work before. From the example that we used before, your routine is going to be getting up from the desk and grabbing a soda when you see someone else with one. The reward in this situation is the soda, so you need to find some other rewards to work with so that you can break the routine.

You don't have to make this complicated. Consider getting some water or some coffee rather than the soda when you get up. This can help you figure out if it is the caffeine or the sugar that the body wants. You could decide to get up and get something to eat. Or maybe the action of just walking around the office a few times can be used as well.

If the latter two are able to work as rewards, this may show you that the drink isn't necessarily what you need, you just need to get up and move around a bit. You may have to experiment a little bit to figure out how the rewards feel for you and which one is the right choice to make you stop picking up a soda. The main goal that you should focus on is to really think through how you feel after you get the new reward. Are you feeling better or worse than before? Are you still fighting an automatic urge to take care of that craving? Evaluating yourself and asking questions will help you determine what rewards work and which ones you should try next.

Isolate your cue

Our brains are going to be responsible for processing a ton of information every minute of every day. Luckily, psychologists have been able to help us out by breaking down the process of isolating our cue. This helps us to figure out exactly what is causing us to do our habit or our routine, and then we can better work at avoiding it or making changes so that we no longer react in such an automatic manner.

First, the moment you start to feel the urge to deal with a craving, you should ask yourself five main questions. These include:

What is the location you are currently at?

What is the time of day?

What does your emotional state look like?

Who is around you right now when the urge   starts?

What was the action that you did right before you had the craving?

We will take a look at each of these questions a bit more later, but being able to answer these will ensure that you are able to isolate what the cue to your habit is. From there, you will be able to make adjustments to better change the habit.

Have a good plan

Once you have taken some time to isolate the cue, at least the one for this problem, and experimented a bit with the rewards, it is time to come up with a plan. You will need to anticipate the cue and then change up your routine to something that is better. This better behavior or routine should be one that will help you to deliver the reward that the mind and the brain are craving at that time.

Chapter 6: Train Your Brain

In the realm of neuroscience and psychology, training your brain uses mnemonic devices. A mnemonic device is a technique that will help you improve your ability to remember something. It's a technique that will help your brain encode and recall information better. A popular mnemonic device is the Loci Method.

Loci Method

The Loci Method translates to location or place, and is based on the idea that you can remember places you're familiar with better than ones you aren't. Therefore, if you can link something that you need to remember to a place that you already know, then the known location will play as a clue to help you remember the lesser known fact.

Here's how the Loci Method works.

You must think of a place you know well, like your house.

Visualize a path you normally take through your house to get from your front door to your back door. Be sure to picture where your furniture is and make sure it's easily remembered things such as your couch, dining room table, and kitchen counter.

Now, as you're proceeding through your house to the back door, place the things you want to remember at each one of those familiar locations of furniture.

When you want to remember that particular thing, visualize your house and go through the same path you took before. The object or thing you want to remember will spring to mind when you get to that location in your home.

Here's an example of the Loci Method.

Suppose you want to memorize the following list of items:

Apples

Canned Cat Food

Nail Polish

Shaving Cream

Cereal

Now, as you're visualizing your house, imagine you're throwing apples at the front door. Don't just imagine the word, imagine the actual apples pounding on the door and creating dents in it as you throw them. Smell the aroma as they burst open and their juices flow down the door. You have to really paint a vivid picture.

Now open your door and enter the foyer or entranceway and imagine giant cans of cat food are rolling down the stairs right at you! You duck out of the way and into the living room.

In the living room you spot a large bottle of nail polish oozing out the goopy color you would like to buy and you know that if you don't get away from it, it's going to drown you in that goopy liquid! Step out of the living room and into the dining room.

In the dining room, you see a bottle of shaving cream in a maid's outfit setting the table. It goes to talk to you and depresses the button on the top, spraying shaving cream all over your face. Don't just imagine it as a picture, but feel the shaving cream sliding down your face and smell it. To get away, you step into the kitchen.

In the kitchen, you see a box of cereal with arms and legs slaving over the hot stove. When you see what it's cooking, it turns you and says, 'Eggs are better.' You see that it's made you eggs so that you don't eat it for breakfast.

All of that may have seemed rather silly, but it's meant to be outrageous so that you can remember the information. You can do this with names, locations, phrases, foreign language words, and so much more. Just be sure that your visualizations are vivid and something you'll remember.

Meditation or Mindfulness (What It Is and Why It's Important)

Meditation and mindfulness are actually two different techniques, but one technique is used by the other. I know that sounds like a bad riddle, but it's true. Mindfulness can be performed anywhere because it is the act of being aware of what you're doing in the present moment. Meditation is the act of using mindfulness while focusing on something such as your breath or a mantra.

So how can you use this in your daily life to make your brain work better?

Throughout the day, you can inwardly pause and become aware of where you are, what you're doing, and how you're feeling. You should do this in a way that doesn't put a judgment label on the experience. So instead of realizing that you're nervous and putting yourself down for feeling that way, simply state inwardly, 'I am nervous'; this will help you realize that what you're feeling is in no way positive or negative. It is simply an emotion.

When mindfulness is incorporated into meditation, it helps the practitioner be aware of their thoughts and yet not react to them. For instance, when you're meditating and a stressful thought comes up about work the next day, but you bring your mind back to what you're doing and the present, that action teaches you how to focus in your day to day life.

So how can you get started with mindfulness and quieting your mind?

Tips and Tricks for Quieting the Mind

Here is one way you can get started with practicing mindfulness right at home.

Start by sitting in a chair or on a cushion with your spine straight. You should be sitting in a comfortable, yet good posture.

Relax with a few, uncontrolled deep breaths and allow your body and mind to relax as you remain alert to your surroundings.

Make a conscious effort to feel the parts of your body that are tense and the areas that are relaxed. Don't try to fix anything at the moment.

Feel the different sensations of sitting and avoid defining or really thinking about it.

Allow yourself to become immersed in the present as you feel the sensations throughout your body.

After some time, bring your awareness to the sounds around you, but not only the active sounds. Bring it to the silence in between those sounds.

Once you're comfortable with that, bring your awareness back to your breaths. Find the part where you seem to focus on your breaths, whether it is your nostrils, throat or your abdomen. Focus on those sensations and how your breaths enter and exit your body, but do not try to control them.

Rest in that sensation of utter relaxation and as soon as you notice your mind is wandering, bring it back to the anchor of your breath without judgment. Do not criticize yourself for allowing your mind to wander because it is a natural cycle of life. Our minds are always thinking.

Once you've mastered that, you can move on to attempting mindfulness while you're at work by focusing on your work and not allowing anything else to interrupt you, or while you're in a social setting such as having a phone conversation. It's not about ignoring everything around you, but being in the present with the other person and really focusing on what they're saying and how they're feeling. Mindfulness will allow you to pinpoint your focus to wherever you want it to be.

Chapter 7: The Importance Of Brain

Fitness

Brain fitness is the healthy state of the mind to handle the various cognitive activities such as reasoning, decision-making, and assimilation of information; comprehension of emotions, motor coordination, memory and more complex process that makes man function every day.

It is an essential factor that you need to make smart and effective response on various stimuli and demands. In order to meet all the things you need to do, the necessity to increase your brainpower is a must.You need to make your brain fit to stimulate quicker connections of neurons and create new neurons that can greatly help your brain to function better.

As you grow older and experienced a lot of mental and emotional challenges, the efficiency of the brain to function to the fullest decreases. This can be attributed to the decline of your physical health that can directly influence the well-being of your brain. Factors like chronic stress that stimulates excess oxytocin and cortisol, estrogen deficiency, emotional traumas and prolonged medications make your brain shrinks its capacity to work well. One of the primary complaints is the memory loss among older people which can be attributed to psychological and biological impairment. It can also be traced by impending risk of Alzheimer's disease and sometimes lack of brain use on high-level skills.

Just like the physical body, your brain can be developed by continuous mental activities like thinking, creative planning, assimilating learning techniques, reasoning power, problem-solving and improving memory. Your brain needs mental stimulation that can make it active and healthy. There is a need to constantly use it to generate growth and development of the neurons in the different regions of the brain especially the frontal lobe where the most vital functions are located like judgment ability, analysis, planning, decision-making and conscious thinking.

When the brain is fit and active, there is an increase of blood flow which brings essential nutrients and oxygen to the cells that works to carry out the different mental tasks.If the brain is healthy, its capacity to bring better results is greater. If the mind is always busy and occupied with mental activities, its productivity increased and protects the brain against degeneration.

It is also important to cultivate healthy lifestyle habits like physical exercise, proper nutrition, enough amount of sleep and stress management in order to keep your brain fit.No matter what your age is, adopting these habits can make you think smarter. The smarter you are, the higher your potential to make wiser decisions to achieve your dreams faster

Chapter 8: The Value Of Meditation --

Simple And Proven Techniques For

Positive Brain Change

We all have, in some form our fashion, heard of meditation on brain activity. As purported in the previous chapters, advanced brain imaging as well as serotonin synthesis adds merit to its effectiveness. For our purposes we willwant to begin with the obvious alterations to brain's gray matter as a result of meditation, and then pose simple meditational practices anyone can use starting today to reap positive brain changes that impact mood, memory, and emotional and behavioral results.

A recent study that was published in the journal Psychological Science revealed that even shorter meditational sessions resulted in noticeable changes in the cortical activity. In fact, a Researcher at the University of Wisconsin discovered that a group of participants assigned to shorter meditation sessions of no more than 25 minutes showed the same changes in the cortical activity as those participating in full sessions.

This finding suggests that even shorter meditation periods can have profound positive increases in emotional activity in the brain.In fact, meditation is just one of those tried and true activities that significantly alter cognitive function and with repetition; the user can develop tremendous positive development of newer neural pathways. Yet again, a study on memory revealed that meditating increases grey matter in hippocampus responsible for memory and learning, while decreasing the matter in amygdala which is the region responsible for the brains alarm systems. Evidence of these physiological changes provides hard evidence pointing to the plasticity of the brain as a result of meditation.

I now present some easy to assimilate meditation techniques that have been practiced and proven to bring about positive changes to the brain, particularly in emotional stability, learning, and memory management. The first technique offered below is very simple and should only be considered a warm up to the more advanced strategies of meditation.

Simple Breath Awareness Meditation

Pick a quiet and comfortable spot where you will not be disturbed

Sit with your spine straight while sitting in a chair with your feet touching the ground

Next, begin to notice your breathing and addcuriosity as to where it goes when it enters your body and how it leaves

Ensure that you do not in any way change the way you breathe, simply breath normally

Quite often you will find that your mind will wonder particularly early on when you begin to practice meditation. Just remember to gently bring your mind back to concentrating on your breathing never being harsh or judgmental in any way.

Continue to concentrate on your breathing for 15 to 20 minutes.

Although this meditational exercise is quite simple, its effectiveness is well established. Yet still, this is only the tip of the iceberg when it comes to meditation. Now the next phase is to impart mindfulness into your meditative practice.

After some practice of focused breathing and gentle redirection of thoughts (as they will stray from time to time) the next step is to develop rewiring your brain.As you learn to notice when your mind drifts while meditating, begin to make it a deliberate mental habit to notice each random thought as it is, recognize it for a moment, accept its pattern, and then gently bring your focus back to breathing.

As you conduct this a few times, you should add some systematic counting as you refocus on your breathing such as counting 1 for the inhale and 2 for the exhale and repeat. You may also put mental words to describe the act such as "I inhale", "I exhale".As you place counts of add words to the act this increases your minds ability to focus on your present state of awareness and develops the strength and accuracy of the neural signals associated with focus.

It is important to remember, however, that you need to accept thoughts and not be harsh or frustrated when this occurs.Simply understand that our minds are made up of thoughts and that we have mental capacity to ponder them, accept each, yet possess the cognition to control them and refocus our attention on anything we desire.

As an added layer to meditation, you may find it useful to observe a single object for at least 5 minutes taking in its shape, color, and feel. You would also want to consider the object for what it is, its purpose, and what life may be like without it. Your chosen object doesn't necessarily need to be anything specific. You could just as easily pick out a random object each day whether outside, inside the house, or at the office.

The point to all of this is simply to engage your brains neural pathways in a way that is different from normal routine and one that is focused on present state awareness.Meditation is probably the fastest and most effective means to achieving brain change.As stated earlier, the evidence of brain chemistry including serotonin creation and growth in axonal and myelin density, as well as enlargement in more sophisticated regions responsible for emotions suggest the value in repetitive meditation. The good thing about all of this is that it doesn't matter if you practice for hours or simply for 5 minutes. You change your brain regardless. Isn't fascinating to know that you truly are not stuck with a stagnant non-growing and age declining brain?The field of Neuroplasticity is fascinating and quite complex; however, reaping the benefits of brain plasticity isn't hard at all.

Chapter 9: What Is Neuroplaticity (Np)?

Neuro Plasticity meaning:
Neuron - means nerve systems that comprise of one hundred billion neurons around our memory gland called amygdala placed on the back side of our brain that often directs 120 trillion of our blood cells, spread across our body, for action. Plasticity - means the way we change our body or facial shape, in order to give our body the most attractive look (mostly celebrities do). Through intense practice, one can transform negative commands of our neurons connected to our sub-conscious brain and get positive outcome.

Our mind receives instructions by five different senses :

1. Visual, 2. Auditory, 3. Kinesthetic, 4. Smell and 5. Test.

1. Visual: All visible matters by our eye come under visual;

2. Auditory: All sound matters by our ears fall in auditory;

3. Kinesthetic: All feelings maters by our skin around our body fall in kinesthetic;

4. Smell: All smell using our nose like sweet smell or flavor, odor, etc. fall in category of smell.

5. Tests: Using our tongue like sweet, sour etc. fall in category of taste;

Our brain mostly perceives and react through the above five senses. For example, if we visited to a party like wedding, birth day, marriage anniversary, etc. we mostly unknowingly come across use of all these five senses to perceive others good or bad reactions towards us. Like others mood of speaking perceived by our ear, viewing all actions by our eye, feeling ones way of touching, smelling the flavor of food, and testing the dishes by our mouth to consider about the quality of party and the friend recorded by our mind. HOW OUR THOUGHTS CHANGE TO OUR BEHAVIOU, THEN TO HABIT AND TO OUR DESTINY?

Internal Representational System :

State of Mind: Thoughts & Moods

Resourceful (Positive)& Un-resourceful (Negative)

Resourceful or Positive state of mind:
Feel Confident, happy, flexible, etc.
Un-resourceful / Negative state of mind:
Feel Depression, anger, illness, frustration, Etc.
1. If we change our actions by programming our internal memory ➡
2. Our state of Mind changes to our behavior ➡

3. Our behavior or attitude become our result ➡

4. Finally our result become our destiny.

How our brain receives data and converts into results?

Thalamus: A part on our brain receives our information from our 5 sensory organs – eye, ear, nose, mouth and skin, and passes to sensory cortex.

Sensory cortex: interpret the received sensory data.

Hippocampus: Not only stores, but also quickly co-relates to similar past events to respond to stimuli.

Amygdala: Quickly interpret emotions and within fraction of a second advice brin for action.

Hippocampus: Releases adrenaline hormone to observe potential threat and decide either to fight or flight.

For example, when we see a tiger on the forest, our brain sense the mood of the tiger. And sense whether it comes towards us to eat us up, then our brain direct us to flight from the scene and save ourselves. Or when our brain sense that the tiger is just pass by the side as its stomachache is already full, our brain direct us not to worry, just stay safe.

Chapter 10: First Books About Trained- Memory Techniques

The Art of Memory by Frances A. Yates
The Art of Memory, published in 1966,is the first academic research book written by Frances A. Yates, a British historian dealt with the history of the mnemonic system starting from the period of Simonides of Ceos. Her works further run through the Renaissance era of Giordano Bruno to Gottfried Leibniz and until the 17th-century developments of scientific methods in the art of memory.She diligently explored the works of various ancient Greeks scholars and explained the development of the art of memory and its metamorphosis. In time immemorial, the art of memory related to religion and the remembering of Holy Scriptures.

The book deals with the thoughts, interpretations and understanding of minds rather than conclusions based on discernible evidence, expressive in nature, text or archaeological.Therefore, it becomes an important thing to learn the art of memory for the people who are into learning and teaching and the memorization process become an important part of the art of memory.

In the Art of Memory, Yates extensively discuss 'memory palace,' which is an imaginary position of the mind where all the information stored temporarily and structured later for processing. Simonides hypothesized the theory of memory palace in the 5th century.In some people, the memory palace will work like to store information in the form of speech.

Simonides accidently invented the theory of memory palace, after a tragic accident, where all the guests perished while attending a function beyond recognition when the auditorium collapsed over them. By remembering the location of the seats where the guests had seated, Simonides was able to identify each of the dead people.He, in fact, was using a technique how he can relate visual image of the 'subject' kept in the brain, to identify the dead.

Yates discussed the convoluted form of memory in her treatise, the Art of Memory and the influence of the thirteenth-century scholar Thomas Aquinas.Her treatise further discussed the contributions of Ramon Lull, Giordano Bruno, and many others in the field of 'the art of memory,' and a memory theater raised by Camilo and the adverse interventions of Church accusing these works as heretical in nature. The responses severely affected the Hermetic thoughts but resurrected as the catalyst for modern psychology.

The Art of Memory, discuss the dominant role of Hermetica in understanding the memory patterns and its contributions to recall information. Despite immense pressure from the Christian clergies, her thoughts had its influence after the Renaissance period, routed to literature and various forms of poetry. Scholars started to accept her views as a rational approach to understanding the complex working of memory, instead of heretical thoughts.

The changes in the attitude were evident in the works of Rene Descartes and Francis Bacon.Hermitica had adopted a more proactive approach towards the views of Robbert Fludd than of Bruno, which might have influenced her in establishing 'Globe Theatre,' the 'memory theater.'Yates has discussed Hermitica's association with Leibniz in length in her treatise, and the efforts made by them to develop a scientific method finding solutions for any issues linking with memory. Unfortunately, the project could not complete due to Leibniz's death.

The book sheds light to the meritorious contributions provided by Hermitica than by Giordano Bruno.In many of the chapters, we can find the extended to study on Giordano Bruno, Hermetic Tradition and about **The Occult Philosophy in the Elizabethan Age** because these treatise highlight interlinks of religion and knowledge in mastering the art of memory.

The book does not provide information about the practices required to train the memory and use mnemonics to remember events by using mechanisms explained in Hermetic thoughts.The book shall give an insight view of the pattern of practices prevailed in the ancient times.Yates has not tried to conclude by saying memorization is an 'art,' or an 'intelligent thought process' present in the brain naturally, instead she left the option open to the readers. We may find it difficult to compare people with trained artistic talent and individuals who had trained to improve the memory recalling power.

The book is still relevant in the 21st century, considering the area of its references, and providing incredible insights into the development of human brain.Even though many of her hypotheses were not acceptable to her contemporary scholars, it seems as she has given too much importance to the occult and esoteric aspects, the book is one of the best-studied academic treatises using by the modern psychologist in academic terms.Notwithstanding all these negative criticism, The Art of Memory will remain as an important paper to understand the development of human memory.

Types of Human Memory

Memory and brain are different.The brain is an organ consisting of soft nervous tissues located inside the skull and memory is the faculty (store) where information is encoded, stored and retrieved as and when required. There are three forms of the information process, and they are **encoding, storage and retrieval**.

Human memory has three main memory sections, and each section has further subsections. Three main memory sections are **sense memory, short-term memory, and long-term memory**.The long-term memory further divided into explicit memory and implicit memory and both of these memories further divided into declarative memory and procedural memory.The declarative memory has two divisions, the episodic memory, and semantic memory.

Further to the above three memory segments, modern psychology has hypothesized the existence of explicit and implicit memory, declarative and procedural memory and retrospective and prospective memory. In addition, we will have the declarative memory again with sub-divisions of episodic and semantic memory.

Sense memory

Sense memory is the micro short-term memory where information degrades rapidly after sensing.Scientific experiments confirmed that this memory segment could hold the information about 200-500 milliseconds after the perception of the object or sound or image or any matter. That information stored for few seconds never transfer to the working memory (short-term memory). In this portion of the memory, the sensory information retained for a short period even after the original stimuli have ceased.Information reaches to this section through 'five senses,' and the five senses are the smell, hear, sight, taste, and touch.

Any information detected or perceived by the senses would disappear immediately if the information ignored consciously at the time of entering into the sensory memory.For this deliberate action, you do not require a conscious effort.The brain will evaluate the information upon sense, and decide on the information whether it may need at a later stage and accordingly process the information.

The sense memory has three types of memory, and they are the iconic memory, echoic memory, and haptic memory.The iconic memory is the short visual information store, echoic memory for short audio information store and haptic memory for data related to touch sense information. In all these memories, it can only store information for a very short period.

The sensory memory can sense the information, hold it for few seconds, and retrieve it immediately. The ability to remember a thing after watching it a while is an obvious example of sensory memory.

Short-Term (Working) Memory

Short-term (Working) memory, as the name denotes stands for the ability of the memory to hold the information for about 10 to 15 seconds or sometimes even up to one minute with an object memorization capacity of about 7 items.This section acts as a scratch pad for storing information and available for processing for immediate actions.This section alternatively called as the 'Post-it Note of the brain.'Short-term (working) memory can remember and process the information simultaneously.

We can explain the mechanism as follows.You can start reading a text, and by holding the information, you can continue reading the remaining.Here the memory holds a temporary piece of the information and in the meantime continues the action for further processing.In this process, the memory keeps the temporary information in the short-term memory, and after finishing, the brain can allow removing the unwanted information.

If the information is essential for you, then you have to apply a conscious effort to hold the information so that the brain can process the information to the next stage of long-term memory for continued retention. You can accomplish it by associating the information with objects or incidents or repeating the information for some time.The processes will encourage the brain to store the information in the long-term memory, and you will be able to retrieve the information when you want to have it.

Working memory and short-term memory are different in concept.Working memory refers to the processes and structures using for briefly storing and using the information, whereas short-term memory is the storage capacity to hold and use the information, in a ready to use format for a short period. Both these terms are interchangeably using, but it differs in terms.

Long-Term Memory

Long-term memory is the segment where information stored for an extended period and available for recall at any time of demand. The capacity of long-term memory is unlimited, and the chances of forgetting phenomena are much less.The long-term memory segment can hold the information for an indefinite period. The information passed into short-term memory, transfers to the long-term memory by a process of collective efforts of reiteration, rehearsal, and meaningful association of matter with images. In long-term memory, the information encodes semantically and stored.At times long-term memory also shows symptoms of acoustic encoding, and that is the reason, we might be able to remember the information by the sound, and even though we are unable to depict, we shall get the feeling that it is 'on the tip of the tongue.'

When in short-term memory, the process of information storage and neuronal communications happens in the prefrontal cortex, frontal executive area and parietal lobes areas of the brain, in long-term memory the storing process happens by using the entire neural connections of the brain. After consolidating the information at the short-term memory, the information takes its root to the long-term memory depending on the nature of information and its utility by the conscious action of the brain.

Chapter 11: Brain Training: Top 20

Powerful Techniques

Now that you have a clear understanding of neuroplasticity and how our brain changes over the years, let's cut to the chase and have a look at all the best techniques to improve critical thinking and memory.

First on our list is Tetris. The classic computer game maybe simpler than most puzzle games these days, but clinical tests have showed that participants who played Tetris daily had an increase in their grey matter of the brain and improved thinking compared to the participants who were not asked to play Tetris. The study showed that playing the game only for 30 minutes daily over a period of three months altered the participants' brain structure. The participants' brain's showed structural changes in the areas associated with critical thinking, movement, reasoning, language and processing.

Creating mind maps help you to improve your memory. It's a great visual tool for you to explore and examine an idea, problem or a task. Whenever, you are studying or learning a new thing, think for a while and mind map. It's a great learning technique that exploits your brain's full capacity. Since, a mind map is never linear, meaning you can't read it in a sequential way, it becomes a scheme containing all the important information. Your brain therefore saves the information in your long term memory instead of the short term memory that is very volatile.

Journey Method is a technique to train your brain into remembering long lists of items. In this memory exercise you have to remember the items in your list based on a journey that you imagine in your head. This is a mnemonic link system where your brain forms new neural paths by associating it with the old ones. So, when you want to remember a list of items, imagine a journey with all the places that you have already visited (the places that are significant to you or you remember them strongly) and associate each of the object in the list with the location of your choice. Since, your brain can readily take up visual information it gets easier for your brain to even remember a long list of items when associated with images and memories of locations you have visited. This method is effective only if you visualize the journey beforehand and have a clear point of reference or landmarks in your imagination.

Pegging is another effective strategy that's used to enhance your brain's memory. If you find yourself remembering through the journey method difficult, then this exercise can also train your memory regions in the brain. Pegging is all about associating known information with the new ones that you want to remember. Here, you mentally place your information with the pegs (information that you know so well, it's almost impossible to forget). For starters, you can use your body parts as pegs, if you have a list of items, peg them to your head, nose, eyes and limbs.

You may also try this method with household objects that you know you are unlikely to forget. With practice, it will get easier for your brain to remember more and more information due to the training.

While pegging takes advantage of your visual memory, you can also use other senses such as smells and sounds to remember things. Similar to visual sense, smell and sound are taken up much faster by our brain. This is the reason you can't forget the smell of lavender easily. You can't forget how the piano sounds. Associating new information with these senses can help you store the information in your permanent memory instead of the volatile one. So, try to remember using your senses. The next time you visit a grocery store, associate your grocery list with their smells and you will have a greater chance at remembering all.

Spatial awareness meaning visual and spatial thinking is a vital skill. When you are driving, you have to navigate from place to place and through busy roads. Even when you are about to plan a picnic, you engage in spatial thinking by planning ahead, imagining how you want everything to flow, how to reach to your destination and etc. All these actions demand you to think visually, arrange and plan. But you can't rely on your day to day activities to enhance your spatial awareness as with repeated activities your brain operates on autopilot mode. In other words, the same actions no longer stimulate your spatial intelligence. In order to enrich your spatial intelligence, you have to practice 3D mechanical puzzles. You can begin with the classic Rubik's cube. You can also resort to visual teaser puzzles. Other methods include learning pottery, carpentry or manipulating computer designs in 3D environment. Even the simple act of assembling furniture from instructions can help train your spatial regions.

Video games are another excellent way to develop visual awareness. Recent studies have shown that engaging in video games can enhance dexterity and short term memory. Video games demand you to spread your attention across the screen to detect and react according to changing events thereby enhancing your spatial attention. In fact, clinical studies have shown that playing video games can actively trigger your previously inactive genes that are crucial to develop neural pathways necessary for spatial attention. Further studies conclude that playing video games can drastically enhance your attention span.

Another technique is to capture your dreams. If you ever find yourself dreaming, try to remember the dream sequence. Grab the moment by writing down each and every detail right after you wake up. You will observe that you are basically describing images of your dreams they might not make any sense at all but dreams are just like that. This way you will start a creative process. This is the kind of activity that will enhance your brain's creativity. By following this method, every time you dream and capture it in your words, you will train your brain to get more creative and imaginative in nature.

In order to train your brain into concentrating effectively, you can try some meditating techniques. First off, find a quiet place, sit comfortably and focus on your breathing. Clear out your mind completely. Don't get distracted by the thoughts running around in your head. There will be moments when you will have a thought flicker across your mind, but try to focus on your breathing and forget everything else. Try to tune out the background noises. Next, think about a place that you have visited. Imagine how it looks, focus on how the weather feels, and engage all yours senses into focusing on that particular location. When you reach that immersive state, you will find that your memory has created a detailed location with its stored memory and filling gaps with its creativity. This approach will force your brain to get more creative and imaginative over time.

Another great technique to enhance your creativity is by playing the game imaginary biography. You can play it alone or with a group of friends. The rules are simple, write down different words on a pieces of paper relating to fame such as celebrities, landmarks or historical events.Fold the paper and put them in a hat. Take turns and pick one paper. Whatever word you get, you have to include it and make up your own biography relating to it for the next 30 seconds. This game is a fun way to train your brain. When you will be struggling to make up your own biography with a completely unknown factor, your brain will have to try harder to fill in the gaps with its imagination and creativity.

Doodle art can be another way for you to train your brain. Scientific studies have found that doodling is one of the earliest ways to children express their creativity. When children doodle at an early age, they might not make much sense at first but with practice their doodles take shape and turn into meaningful drawings. Furthermore, research has shown that doodling actually helps you to organize your thoughts, feelings and experiences. If you ever have a creative block, you can always resort to doodling. It's a way for

your brain to express itself in form of ar.
when it fails to answer through words. So,
to train your brain, you can try and
interpret doodles by others. If you can't
think of anything then add to the scribbles
and keep on adding till you can turn it into
something meaningful.

Lateral thinking is quite important,
especially when we want to solve a
problem but can't seem to find any
solution. Lateral thinking is all about
having a different perception instead of
following a linear path while creative
thinking. We tend to follow the way that
others have followed. We overuse logic
and limit ourselves within it. The result?
We don't find a solution at all. Because,
digging a deeper hole in the same spot will
not change the outcome, digging the hole
some other place will. So, here's
something that you can do to enhance
your lateral thinking and broaden your
perception. Think and answer, how many
different ways can you think of to join all 9
dots in a square using only 4 straight lines
without taking off your pen of the paper?

Take a paper and try all the ways you can join them. This will train your brain to think out of the box. You can try similar exercises and puzzles that demands you to think out of the box instead of following the trend.

Optical illusions are another effective way to train your brain. Optical illusions give an insight on how your brain's creativity works. The study of visual illusions and mental fallacies has proven that eyes doesn't really see but collect information for the brain to process. It's the brain which associates the collected information to the context, analyzes it, processes it and qualifies it. Believe it or not but making sense of an image is one of the most creative act that our brains are capable of.Optical illusions are known as great creative stimulators because they engage the right hemisphere of your brain (associated with creativity) and forces your brain to perceive things differently in an attempt to interpret the image. As exercise, you can try looking at different kinds of optical illusions and interpret them.

Next on the list is "Number Workout". Just as a weight session in the gym helps you build muscles, the number workout exercises your brain to enhance speed and IQ. The number workout is a very simple numerical exercise. It will often involve basic quick fire arithmetic, simple mathematical problems such as additions, subtraction, multiplication and division. At first glance, this exercise might seem like nothing special but there's to more to mathematical exercises than meets the eye. Clinical studies have shown that the nervous system of your brain contains neurons, axons and nerve fibers, all of which are associated with the transmission of information impulses to and fro. The speed at which these impulses travel between your brain's neurons determine how fast your brain can process the information. And you will be surprised to know that solving a basic mathematical problem like addition increases the insulation around axons which results in faster transmission of impulses across the neurons. Furthermore, Research shows that mental arithmetic helps speed and accuracy and more complicated and elaborate math problems like trigonometry or algebra boosts your problem solving ability.The problem with us is, we rely on calculators to do these calculations for us, but you need to engage your brain in order to keep it in shape.

You can also try visual math workout as well to increase your smartness. These exercises help your brain to process information faster thereby increasing your smartness. In other words, a problem that could take others longer to solve, you will be able to solve it faster. You can begin with simple geometric problems with diagrams, or solve problems that involve pie charts and bar graphs. These are excellent exercises to work out your brain. When you try to solve visual math problems, different parts of your brain become active and engage both the right and left hemisphere. You can interpret the mathematical language by finding ways to visualize the logical meaning. With visual math problems, it may seem daunting at first as too much information is provided and you could feel confused. But with more information it can only get easier to solve a problem compared to less information. Moreover, studying the visual diagram of the math problems can make it much easier to solve the problem.

If you want to enhance your logical thinking then Sudoku is by far your best choice. Over the last few years, Sudoku has become quite popular for exercising your brain. It was originally popularized in Japan during mid-1980s. However, it was initially known as the "Number Place" and the creator of this game was an American named Howard Garns. It's a neat little puzzle, with 9 squares placed to form one big square. Sudoku is basically a logic exercise. It primarily uses numbers but involves no arithmetic. Playing Sudoku on a daily basis along with other puzzles and exercises will help keep your brain active. Our brain ages with time and deteriorates faster because our work and lifestyle is often linear. As we do things repeatedly, our brain switches on its auto pilot mode which doesn't stimulate the brain as much. Playing Sudoku will really bring a change into your routine and engage the brain to work and enhance its logical thinking.

Riddles are brain teasers that help your brain to improve its critical thinking. Riddles can be a fun way for you to engage yourself in brain training. These are very similar to logical fallacies that make you employ false logic. Since they are expressed in metaphorical or allegorical language, they are very obscure to understand at first. But that's the main essence of riddles, they are designed to make you confused so that you have to think harder and concentrate more to find out the correct answer.

General intelligence comprises of two components: fluid intelligence and crystallized intelligence. Fluid intelligence is the ability to find meaning in chaos and find out solutions while crystallized intelligence is about all the knowledge, information and experiences that you accumulate over your lifetime. Therefore, when children are young they learn language faster and over time it becomes crystallized intelligence and doesn't really spark up your neurons much. But if you learn a new language in your adulthood, you will make your brain active as learning a new language is demanding activity. It takes up a lot of resources of your brain but also makes your brain stronger in the process. In fact, clinical studies have shown that learning a new language is one of the best ways to protect the brain against the ravages of aging. In a study, participants who were bilingual suffered less mental from aging compared to the participants who spoke only one language.

Who knew you could get smarter reading and interpreting comprehensions! As a matter of fact, reading comprehension does work your brain in multiple ways. The ability to understand and interact with the words in the comprehension works out your perception skills, reasoning, problem solving and other cognitive functions. You can begin with reading a few passages from books with comprehension exercises and then answer the series of questions related to the passage. This approach trains your brain to draw out logical interferences from simple life situations. Your answers are usually a reflection of how well you can interpret the comprehension. Plus, it will also engage your visual memory, short term memory and speed at which your brain is capable of transferring impulses over visual pathways.

Neuroplasticity experts have also found out that musical training improves the brain's overall function and connectivity. In fact, a study in 1993 showed that participants performed better on spatial reasoning test after listening to Mozart sonata. In order to train your brain, you can both listen to music or learn to play a new musical instrument. When you learn to play a new musical instrument, your brain undergoes major changes. Studies have shown that there's a drastic improvement in the brain's speech perception, the ability to interpret emotions and multitask simultaneously. Learning to play music forces your brain to get better at problem solving skills, switching between tasks and concentrating. Since, brain plasticity results from experiences that engage the brain and places higher demand on it, learning to play a musical instrument is a great tool to bring about positive changes in the brain.

Chapter 12: Not So Fast

The same traits of neuroplasticity that make your brain resilient make it vulnerable to negative outside and internal (usually unconscious) negative influences. With a harsh living environment, less than stellar upbringing, or genetic mental illness, neuroplasticity can result in over-reactive, depressive, anxious and obsessive patterns.

The Science of Neuroplasticity

So, what's behind the curtain of this revelation about our brain's resiliency? We intend to deliver an understandable definition of what's going on in your brain when it changes. Being aware of what is happening in your brain is the first step to making positive changes that will result in positive outcomes. You should be excited about this. You can have the future you desire! It's available to you and you have a brain that is designed to take you there.

Your brain is made up of up to 100 billion or so neurons that make up to 10,000 connections with other neurons. (These connections are also called synapses). Remember this part about connected neurons – this piece of information is very important to get under the hood of neuroplasticity.

We are talking about a process where the brain's neural synapses (connections) and pathways are altered by behavioral, environmental, and finally, neural changes.

Fun Facts About Neurons

Neurons talk to each other by sending chemical signals across the synapse (the connection between the neurons)

Neural impulses are an important facet of how the brain works – neural impulses code your actions, thoughts, and experiences.

These thoughts, actions, and experiences reorganize the structure and function of the pathways that neurons use to talk to each other. This enables the brain to respond to the world around you.

Neurotransmitter release is also determined by patterns of neural impulses.

A neurotransmitter is a chemical released from a nerve cell. This chemical then sends a message from a nerve cell to another one, or to a muscle, organ or tissue.

Basically, a neurotransmitter is a message from one cell to another.

Amazing stuff, right? Here's an example: Neurons send signals across connections. These result in a thought. The thought changes the way that your neurons talk to each other, so you can respond accordingly. If you have a negative thought, a neurotransmitter is released from a nerve cell, sending a message to the muscles in your face, telling them to frown.

Still not clear? Don't worry, by the end of this book, you will be an expert on the inner workings of neuroplasticity.

How Experiences Change Your Brain

This is Your Brain on Neuroplasticity

Three Analogies

Here are three well-known analogies to put the concept of neuroplasticity into perspective.

The film analogy: Think of your brain like blank film. Say you take a picture of a scenic landscape. You are exposing this film to new information. The film reacts to the light it is exposed to, and its makeup changes so that it can record the image of that landscape. Just like film, when your brain is exposed to new information, it changes, so that it can retain the information.

The lump of clay analogy: Say you take a lump of clay and press a coin into it to make an impression. For the impression to be visible, changes must happen in the clay. The shape of the clay morphs as the coin is pressed into it. It reorganizes its makeup as the coin presses down. Just like the clay reorganizes its makeup, neural circuitry in the brain reorganizes, too, in response to new sensory stimulation or new experiences.

The music analogy: Imagine that you just started learning how to play the flute.

Learning how to play a musical instrument demands a lot of cognitive, sensory, and motor skills. Think about it – you are reading music and translating what you read into ambidextrous movements that depend on auditory feedback.

With practice, you will develop fine motor skills, nail down time signatures to develop precise timing, and eventually memorize long compositions, among other challenges. This enhances neuroplasticity big time – many levels of the brain are being used at the same time.

When you first picked up the flute, you made mistakes and were very clumsy. Flute players will tell you even eliciting a sound from the instrument can be a challenge for beginners. Yet only within a few days, neural circuits in your brain devoted to fingering your keys will begin to fire repeatedly.

The more your neurons fire, the stronger the synaptic connections become. Think of neurons as friends – the more they communicate, the stronger the friendship and connection is. With practice comes proficiency. Neurons in circuits that are involved with recognizing music tone will connect more distant neurons because different parts of your brain need to speak to each other now. (Including both hemispheres!) The more connections between different areas of your brain, the better.

Your Brain is More Plastic During Childhood

During childhood, your neuroplasticity switch is always on. In adults, it is typically set to off. Don't despair if you are an adult – specific conditions can trigger or enable plasticity, turning the dial back on. You CAN grow your mind as an adult, and the steps you take to tap into plasticity can improve your personal, spiritual, and work life, as well as your health.

Steps like focused attention, hard work, determination and maintaining brain health will kick plasticity into high gear. Lifestyle changes that are good for the body, like exercise and diet also benefit the brain. You will see as you read on that your brain takes cues from your body and your body takes cues from your brain.

The best news? Keeping your brain stimulated might delay the onset of common brain diseases like Alzheimer's and dementia.

Chapter 13: Some Of The Most Common

Mental Models

The neat thing about these mental models is that there are actually quite a few of them that you are able to work with. This allows you to think and act in many different manners and can help you when it is time to make some big decisions. With this in mind, we are going to focus on the top ten mental models, even though there are at least 80 of these models that can be applicable to our modern world. The top ten mental models that we are going to focus on in this chapter will include:

The Map is Not the Territory

The rationale that we are going to follow with this model is that the way we are able to see the world isn't really itself. Rather, the worldview that we have is going to be based on our own mental construct. A map is just going to be the mental construct that we use. Territorial borders are going to change over time, and things around us can change as well.

In a greater perspective, the map is not the territory will warn us against some of the logical fallacy that occurs when we confuse the labels, semantics, and artifacts with the things that are real. Some of the other fallacies that are similar to this will include misplaced concreteness and reification. Each one of us is going to have or own mental map that helps us to view the world. Yet the world is really complex, and it is impossible for a human to totally comprehend what is going on. This means that our attitudes, beliefs, assumptions, and conclusions are not going to be the best parameters all of the time for helping us to understand the totality of the real world.

When it is time to look at our communication and interpersonal relationships, it is likely that we are going to try and impose our own mental map onto others, or we think that they should be able to read our mental map. But we have to remember that the other people we are around have their own mental maps as well, and they would like us to follow theirs instead of ours. This is where a lot of misunderstanding, confusion, and sometimes extreme conflict, is going to come into the picture.

Yes, it is possible for two people to be in the same place, and even see the same situation or circumstance, but then leave with a different experience. With these different experiences, we are going to see that there are different interpretations and different worldviews in the process. If this is the mental model that you are going with, remember that you should not try to impose your worldview on others, just try to make yourself understood, while also accepting the worldviews of others.

There are several ways that you are able to apply these mental models in your life. These include:

Decision making

Relationship management

Leadership

The Circle of Competence

There is no one who has all of the knowledge in the world. There are always going to be some blind spots in our competencies and the knowledge that we have. it is important that we acknowledge the things that we don't know. This is not a bad thing to not know everything, because it is just not possible. Focusing on the things that we do know, and seeing how that can move us ahead in life is so much better.

Through the circle of competencies, you are going to be able to open up your mind to some more learning. You will be able to avoid some of the common assumptions and fallacies along the way. You can work to discard the ignorant ego and become more reasonable in your understanding the extent of the competency that you have.

This is going to benefit you so much. It can help you to know where your strengths are, and your weaknesses. It is able to point you in the direction that you need to break away from the cocoon of mediocrity and will help you to connect with the right kinds of people so that you are able to learn more from them and grow your knowledge base. Some of the different applications that you can use with the circle of competence will include:

Career development and advancement

Decision making

First Principles Thinking

It is going to be a lot more about separating the underlying facts and the ideas from the assumptions that you are able to make. when you decide to apply the first principle of thinking, you will be able to decompose a problem into its constituent elements, or to the root causes that are present, and then it is easier to deal with them in the most appropriate manner, rather than running around and hoping that you make the right decisions along the way.

You are able to isolate out some of the pathogens that will cause you the disease from its symptoms. And because of this, you will end up providing some healing to the disease, rather than just some relief to the symptoms that you are feeling. Reaching to the depth of the constituent elements, you will find that you can use these elements to build up to something that is brand new. Some of the applications that you are going to see with this one include:

Decision making

Problem-solving

Engineering

Chemistry

Medicine

The Pursuit of Knowledge That is Liquid

Solid knowledge is going to be comprised of pellets that are going to be collected into silos. For example, you could have biology silo, a physics, silo, mathematical silos and so on and solid knowledge is hardly going to be fluid, hardly dynamic at all, and hardly flows.

Great free thinkers are going to go beyond the solid knowledge that is offered in schools to the liquid knowledge that is uncondensed, unrefined, and freely flowing. This kind of thinking is going to be a knowledge gathered from your own experiences, from your own discoveries, and your own exploration. This allows you to have a nice adventure while you work on your discovery of knowledge, which can make things easier and more pleasurable along the way.

Thought Experiment

Thought experiments are going to be devices of the imagination that are then used in order to investigate the nature of the things around us. Thought experiments are going to be essential any time that you would like to break into new frontiers, especially when you are trying to go into new and unknown territory. This territory can be unknown and new to you, or they can be unknown to others as well. This kind of thought experiment is going to allow one to crack the impossible, evaluate the potential consequences, and then compare these consequences with some of the known to help them make some informed deductions.

You will find that, in a manner different than some of the empirical examinations that you can do, thought experiment is going to be conducted just in the mind, even though you are able to demonstrate it physically in order to prove it. As such, this kind of mental model is going to be known as a laboratory of the mind instead. Galileo is actually one of the famous scientists who worked with this thought experiment to help come up with some of the scientific principles that are still seen to be true today. Albert Einstein also used this method as well. The biggest challenge that comes with this kind of mental model is that, without being able to prove it empirically, it is something that we are not able to prove as true or false. Thus, there are some philosophers who are going to consider it more of a mental modeling of the physical realm instead. Some of the applications of this mental model will include:

It can help you to research the formulation of your hypothesis.

Scenario simulation and synthesis.

Scientific exploration
Second-Order Thinking

The first question that we may have here is whether or not there is a First Order? Of course, there is! Before we start to take a look at this Second Order of thinking, we need to do a brief introduction to what the First Order is all about and then compare the two.

When we are working with the First Order thinking, people are going to make decisions that are hasty and quick, ones that are based on what they can see on the surface and in appearances. Because of this quick decision making, the person is not going to be able to go into the depth that they should to understand why things turn out the way that they are. They react quickly, on first impressions, without actually focusing on what is happening and whether they should react in this manner or not.

Then we have the Second Order thinking. With this one, the person is going to take their thinking and decision-maker a bit further. They are going to study some more of the fundamentals that are behind the phenomenon so that they can identify the various variables that are behind this as well.

Those who end up using the Second order thinking, rather than rushing into the decision and going into the decision too quickly, may find that they make a decision that contradicts what their First Order thinking would have done. Let's take a look at an example of how to do this. When there is an explosion that happens, the First Order thinkers are going to try to run away because they assume that the explosion was from a bomb and they want to get away from the danger as quickly as possible.

But then there are the Second Order thinkers. Even though they may want to follow their instincts and take flight, they will be more likely to pause and think about what it could be that caused the explosion, figure out if they are close to the explosion, and then figure out what they could do about it. Maybe they find out that it was just a big tire that burst by them and there was no reason to run off and be scared of it.

So, in some manners of thinking, the First Order thinkers are going to reach to the already existing mental image that is in the mind of this kind of thinker. The actual occurrence that shows up is going to just be the trigger for that action, but not the cause of the action.

Another good example of First Order to Second Order thinking is in the sphere of investment. For example, if Company A declares that there is a profit warning, most First Order thinkers are going to anticipate that the price of the share is about to fall. Because of this, they are going to rush to get rid of the shares as quickly as possible. This causes the shares to happen, but this is because the traders jump in too quickly, not because the value of the shares actually went down.

On the other hand, we can see that a Second Order thinker is going to have done the process of a fundamental analysis on the profit warning to figure out what is going to cause that profit warning.

This could easily be something small, like a new investment decision that resulted in less profit, but a bigger asset base. This means that if you hold onto the shares, there may be a temporary dip thanks to the First Order thinkers, but overall, your value from the shares are going to heat back up.

There are a lot of different times when you will be able to use the idea of First Order vs. Second Order mental model to help you make decisions. Some of the best applications of this mental model are going to include:

Decision making of any kind

In a fundamental analysis

Investment decisions

Emergency response

Occam's Razor

This is a model that is going to posit that the simple explanations are the ones that are more likely to be true, rather than going for the explanations that are more complex. So, if you are using this mental model, you will find that it works the best if you can pick out a solution that has fewer assumptions in it.

What this means is that when someone makes a decision, they need to be able to minimize the assumptions as much as possible, with the help of experimentation, study, and research before they implement a new decision concept that they want. In case there is not a lot of leeway in the amount of time that is available, then the concept that has the fewest assumptions is the one that is considered the most ideal for implementing.

The reason for this is that when you have more assumptions present, the risk of having some decision errors is going to be higher. Some of the applications of this mental model is going to include:

Personal development

Career choices

Management

Leadership

Decision making

Inversion

The next mental model that we are going to take a look at is inversion. Inversion is the idea that you will think backward. With this one, you are going to create some likely scenarios after the action or decisions and then you need to seek out how to address all of the scenarios before you decide on which course to take, or before you decide to execute it. This means that you need to be able to approach the problem from the opposite of the natural starting point.

Let's take a look at an example of this. You may be considering a separation or a divorce from your spouse. Before you jump into this and go ahead with some of the divorce proceedings, by inversion, you would stimulate some of the potential scenarios that would happen with this. It could include a strained relationship with your in-laws, loss of your home, increased costs of childcare, loss of your partner, property division and more. If possible, you can then start to mitigate some of the adverse effects of these scenarios before deciding for or against the divorce.

To make this one work a bit better, you have to make sure that you attack your decisions by going backward. This helps you to really plan things out, imagine that things are already done, and consider what decision is actually going to give you the results that you are looking for.

Some of the different applications for the inversion mental model will include the following:

Career choice

Leadership

Decision making

Personal development

Family planning

Probabilistic Thinking

When you work with probabilistic thinking, you are going to use a variety of probability tools, including statistical tools, to help them approximate the likelihood that a certain event is going to occur. There are two major technologies that already use this including machine learning and artificial intelligence. Some of the different applications of using probabilistic thinking as a mental model will include:

Strategic planning

Actuarial science

Computer sciences, especially when we look at machine learning and artificial intelligence

Decision making

Investment

Hanlon's Razor

The best look at what the Hanlon mental model is all about is "never attribute malice to that which can be simply explained by stupidity". The gist of this is that you should never assume that something bad is happening because of the wicked intents of others on the situation. Sometimes it is just incompetence or stupidity on the part of the actor, rather than some malice on their part. Stupidity, in this case, can be from the other person, or from you. It is possible that your own stupidity is causing problems because you made the wrong assumption, but then it could be the stupidity of the other person as well.

Due to the egocentric view that most of us have of things, which, in our subconscious minds, assumes that everything in the world revolves around us, we end up assuming a prominent role in the story of everyone else, even though this isn't true. This means that when we are around someone who seems a bit annoyed, we assume that it has to do with us and that we have made them made. When another person is rude to us, it is because they are angry at us or just being mean to us. When we see that someone doesn't want to congratulate us, we assume that they are feeling jealous of us.

As part of the stupidity that comes with us, we are going to perceive some negative responses as malice against us, without being able to consider that it is often due to factors that have nothing to do with us. Hanlon's Razor model is going to be helpful because it can avoid paranoia, anxiety, and stress. It will eventually save us from taking a bad situation and making it worse.

When we use this mental model, we have to understand that to be human is to err, and that there are times when people make mistakes, and even we are going to make some mistakes on occasion. This means that we need to find out if there are other explanations for what has occurred, rather than assuming that there is some malice that comes with this. This is why with Hanlon's razor, it is always best to assume the best intentions or some good faith before we try to prove otherwise.

Some of the applications that we are going to see with the Hanlon's razor will include:

Relationship management

Decision making

Diplomacy

Conflict resolution

Crisis management

As you can see, there are a lot of different mental models out there that will help you to make decisions, based on the way that you see the world and the point of view that you are looking for along the way as well. Each of these can be effective and ill help you to see some of the results that you want with making decisions that are going to be based on sound judgment, rather than on our emotions or something else that can be subjective. You can determine which of these mental models, as well as some of the others we will talk about in this guidebook that you would like to use to help you make some good decisions and to ensure you can make the best decisions for yourself.

Chapter 14: Neuroplasticity Exercises

Contrary to what some people have thought, Neuroplasticity is not something found only in one single structure of the brain. Equally as important to understand is that it does not consist of one type of chemical (or physical) event. Instead, the ability to mold the brain, or its plasticity, is the result of lots of different and complex processes, processes that occur in the brain throughout our entire lifetime.

One of the hottest subjects in the media these days is the brain, and the various effects we can have on brain development. We also see that brain health is one of the fundamental requirements to the health of our body in general, so the introduction of, or continuing focus on things like Neuroplasticity exercises, is fast becoming a subject of interest.

We exercise virtually every part of our body every day and yet we don't very often place any real emphasis on exercising our brain. Sure we all think, and use our memory, but this in itself is not considered focused brain exercise.

Providing your brain with the stimulation known as Neuroplasticity exercises, (brain gym) is one way of formulating a routine that will provide the brain with a tactical approach to development, and increasing the ability of the brain to do more. If you stop and consider for a moment, that everything we do or think or say, smell and hear, is controlled by millions of neurons in the brain, it makes sense to be able to manipulate the brain to be able to control even more.

Neuroplasticity exercises give us reason to believe, as some scientists do already, that by regular and frequent use of certain activities, the brain and its function can only get stronger. Each memory we have for example, is believed to be proven to be an electrochemical process configured by and entangled in the neuronal circuitry inside our brain. Without regular and frequent exercising, these electrochemical units can actually go into a state of hibernation.

So, by gently persuading this Neuroplasticity, by using the correct activities we know that we can indeed improve and enhance brain function. By being able to manipulate the brain through the use of Neuroplasticity exercises, we can even start to understand, and fend off, symptoms and ailments associated with brain disease, like Alzheimer's Disease, and of course absentmindedness.

I think we have all heard the reference to the theory that we only ever use (in normal situations) around 10% of our total brain capacity, and not as many people would know that we can actually enhance this to around 50% by implementing and following a regime of simple games, known as Brain gym, or Neuroplasticity exercises. By following and playing these games, we are able to stimulate the mind, and increase the levels of working memory, enhancing attention span, and speeding up the processing power of the brain.

Increase your own brain function quickly and easily by implementing the Neuroplasticity exercises whenever you can. Brain Gym exercising is as important to our well being as everyday health care.

FACT

What are neuroplasticity exercises? Perhaps it would be helpful to know what neuroplasticity is.

Neuroscientific dogma up until about ten years ago was that past a certain point in our lives, our brain did not change, we had what we had, and that was it. In fact, any change would be that it shrank, actually, as neurons died.

But that dogma has been overturned. Our brains are constantly trying out new connections. Each neuron has branches which are connecting with other neurons based on what we are learning in the moment.

Those connections can be kept, and form what is called a cognitive reserve, which can reroute signals around trouble spots in an aging brain, for example. Here is how neuroplasticity is described in a review of Sharon Begley's book, Train Your Mind, Change Your Brain, from the Mindfulness Institute.

"For decades, the conventional wisdom of neuroscience held that the hardware of the brain is fixed and immutable ? that we are stuck with what we were born with. As Begley shows, however, recent pioneering experiments in neuroplasticity, a new science that investigates whether and how the brain can undergo wholesale change, reveal that the brain is capable not only of altering its structure but also of generating new neurons, even into old age. The brain can adapt, heal, renew itself after trauma, and compensate for disability."

"Begley documents how this fundamental paradigm shift is transforming both our understanding of the human mind and our approach to deep-seated emotional, cognitive, and behavioral problems. These breakthroughs show that it is possible to reset our happiness meter, regain the use of limbs disabled by stroke, train the mind to break cycles of depression and OCD, and reverse age-related changes in the brain. They also suggest that it is possible to teach and learn compassion, a key step in the Dalai Lama's quest for a more peaceful world. But as we learn from studies performed on Buddhist monks, an important component in changing the brain is to tap the power of mind and, if particular, focused attention. This is the classic Buddhist practice of mindfulness, a technique that has become popular in the West and that is immediately available to everyone."

In a Blog Talk Radio interview with Simon Evans, Ph.D., co-author with Paul Burghardt, Ph.D. of Brainfit for Life, Professor Evans reports that those new connections can happen in moments, perhaps hours.

In other words, I do not have to attend 50 or 100 lectures to reach a knowledge tipping point, and then suddenly I have new connections.

Those connections are part of the ceaseless activity already going on in my brain.

Begley's work focuses on contemplative approaches to neuroplasticity, or the use of meditation, such as that practiced by Buddhist monks.

In the west, research is, or course, revealing some technological tools for enhancing neuroplasticity exercises.

In particular, Micheal Merzenich, Ph.D. of Posit Science is demonstrating some very strong results using the auditory training in the Brain Fitness Program.

I happen to own and use this program, and have anecdotal successes to report from its use.

I have also been a student of Chi Gong for about 9.5 years, and find that I can focus for much longer periods of time than I used to, especially when exercising.

So it appears that there are a number of neuroplasticity exercises we can participate in, either of the technological or contemplative traditions.

Chapter 15: Neuroplasticity And Memory

Your memory is incredible. There is seemingly limitless storage, and sometimes it can surprise us. Have you ever just randomly been sitting there and thought of something that happened over a decade ago? Do you wonder why it's so easy to remember certain meaningless information but hard to retain the stuff that matters? Throughout this chapter, we will teach you all you need to know about your memory.

Why We Forget Things

Before we start discussing the ways that you can improve your memory, let's first look at the reasons why you forget things in the first place. The primary goal is that as you are learning the new information, your brain doesn't even think that it's important enough to keep around.

As you learn new things, you start storing these experiences, skills, and knowledge in your short-term memory.

If your brain thinks that it's important enough to keep around, then it'll move these new thoughts into your long term memory. Your short term memory will include the quick visuals that you see little bits of info that you hear and other small things that you pick up on. The long term memory will be a generalized idea of the situation that happened in that long term memory of your story, some of the small stuff, but a lot of it will get lost in the process. The first reason that you might not entirely be remembering things is because they're not going to your long term memory in the first place.

If your brain isn't storing things properly, then you won't remember things correctly. To remember things better as you're learning them, you can do a few things. You can make sure that you put all of your focus and concentration on memory and keeping these bits of information stored. If, as you're learning something, you make sure to emphasize remembering it, then it will stick around easier.

Another reason that you might have trouble remembering things is simply that there's interference, distraction, and other things that are making it hard for you to put your full concentration on what you need to learn.

If there are noises around other visuals, people talking stressful thoughts and anxious fantasies rolling through your head constantly, it will be a lot harder to remember individual bits of information. We also have to consider the way that we might be subconsciously suppressing or disassociating at the moment.

Think of the last time that you were stressed out while you're having a conversation with somebody. Maybe you were thinking about all the work that you had to do, the long list of tasks that needed to get done that night, and everything else that was making you feel completely stressed out.

All of these thoughts flowed through your brain while somebody else was trying to talk to you. Maybe they were going over plans for the weekend, telling you a story, or giving you something else that you probably should have remembered. Later on in the week, they asked you about that story, and you can't remember a thing because you were too stressed out to listen in the first place. Sometimes we disassociate from the moment because of our anxiety.

In order to really start to improve our memory, we need to focus. We need to learn how to pay attention and stay present in the moment. Before we give you tips on that, let's remember to consider the incredible storage that our brain has.

Your Brain's Storage

Sometimes it can feel as though we are running out of space to keep things stored in our memory. What we have to remember is that our brain is not like our closet.

You could fill your closet with a ton of boxes but eventually might run out of space. Your brain is pretty much never going to run out of storage; you will always have extra storage to use in your brain.

Any phone computer, tablet or other electronic device that you have ever had still does not compare to the amount of storage that you have in your mind. You could probably read every book that's ever been written and yet have enough room in your brain to learn even more.

The reason that you remember things will never be because your brain is filled with storage. Sometimes you might wonder why you remember the jingle of a commercial you saw when you were six years old better than you remember your boyfriend's mother's birthday.

But it's not a matter of storage, and it's not as though you need to clear out space in your brain. Your brain would do that on its own, anyway.

Your memory will never be affected by the things that you already know. Your mind will be affected by what you do as you are learning that new information. Afterward, it's important to remind yourself of all the new things you learn so that you don't forget them.

Your brain has seemingly unlimited storage so never assume that this is going to be the reason that you can't correctly learn things. Your brain can always stretch, can always change, and can always modify shapes. It is not like a hard drive, where there's only a certain amount of things that it can do.

Your memory is plastic, and it's time to learn how to use it to its full benefits better.

How Depression and Anxiety Affect Memory

Those who suffer from anxiety and depression probably already understand how hard it can be to remember certain things. Let's look at a few reasons why these mental illnesses can cause you to have less of an active memory than other people.

The first reason that anxiety and depression might affect your memory is because of your fight or flight response. When you get into that kind of ready state that a fight or flight response triggers your brain is going to stop working as hard to remember things, all of your energy is going to go to protect you. Think of the last time that you were under an intense amount of fear. You probably don't remember it; there's a good chance that you have some traumatic experiences that you really only remember from somebody else's perspective. That is because your brain has stopped worrying about picking up on information and is putting all of its time focus and energy into making sure that you stay alert to protect yourself as best as possible.

Another thing to consider is that sometimes our brain will shut out certain memories for our own protection as well. If you are always depressed, you might be in a constant state of dissociation because it is easier for your brain to manage this than it would be to manage the awful feelings and emotions that you have while depressed.

Think of how sometimes if you see something you don't like, maybe a news story, a gross video, or anything else like that, you just look away out of instinct. That's almost what your brain does when you are anxious and depressed. It doesn't want to have to experience those hard emotions. If you experience those negative emotions, it causes you to feel bad. And that has adverse effects on your body. As a defense mechanism against these mental illnesses, you disassociate so that your brain no longer has to deal with these kinds of things.

What we have to remember is that having anxiety and depression doesn't mean that you won't be able to remember things. We just have to put more of a conscious effort into remembering these things. In fact, anxiety could even make your memory better when you're anxious; it's easier to pick up on some smaller details that other people might not see. For example, let's say that you're really anxious about your appearance, you might pick out all the little small details about your face, your outfit, your hair, and all the same things and other people as well.

What you have to do is just remember those things, remind yourself of those small details. Make little notes in your brain to keep track of the small things you pick up on while in an anxious state.

Sometimes these things can set us back. But if we are stronger than them and know how to navigate with these mental illnesses, it will only help us in the end.

Decluttering Your Brain

As we already discussed, your brain will never run out of storage, and it's not like your closet where you have to clean old things out in order to make room for the new. However, it can still help to declutter your mind and clear things out in order to make your brain function more optimally later on. Let's go over a few steps that you can start to declutter your mind in order to have stronger neuroplasticity. The first thing you can do is to declutter your physical space. You might not realize how important this is because it's a subconscious thing that happens. But your brain will still process the things that it sees in front of it. Let's say that you're sitting at your desk right now in front of you, you have a computer, a little cup with a bunch of pens in it, a stack of paper, a picture of your friends, a plant a lamp, something else hanging on the wall, a few other things on your desk decorations here and there.

At the same time, as you're sitting at your desk working, you can still see every little thing out of the corner of your eye. You don't sit there and think plant, cup, mail, and so on in your head, but your brain still sees it and picks up on it like a camera. You're again using a little bit of mental energy to process these things. For most of us, it's not enough to where it's distracting. So, in a shared space, little decorations here and there will not hurt you. However, if your area is cluttered with unfinished projects piles of trash and things that distract you, then this will take up some of your brain's power.

For example, let's say that you have a half-painted chair in the room that you want to finish. Maybe there's a quilt that needs to be sewn. Perhaps there's a whole entire dresser that you need to cleanout. If you can see the side of the corner of your eye when you're working your brain is still going to pick up on that. It's always going to register the fact that you need to do something about this as well. It's going to remind you that there's a task that needs to be completed.

How to Recall Information

Once you learn how to remember things better, that can really change your brain. However, what can you do in order to recall that information that you're learning?

Sometimes we take in information, and we do know how to store it properly. But when it comes time to recall it, it's tough to do so. How many times have you had a conversation with somebody, and maybe they asked you to bring up an example of what you're talking about. You have a million cases in your head to use, but you can't recall any of it for some reason. There are a few things we can do to make recall easier. The first thing you can do is use an anchor. An anchor something that will keep your mind focused on the moment that you're trying to recall.

Let's say that you're learning the names of your classmates.

Five names are Judy, Ben, Kyle, Tom, and Jessica. Each time you look at their name, you could touch one of your fingers to the table. Say, Judy, contact your thumb to the table. Say, Ben, move your pointer finger to the table and so on. Say, Kyle, touch your middle finger to the table, and so on. When you can't recall their names, try to mimic that touch, and you will discover that you can better remember the names you've learned.

Another way to better remember things is with mnemonics. This is taking the first letter of each word and associating it with something else can be like an acronym, even.

For example, think of the program. D.A.R.E. This stands for drug abuse resistance education. However, as a whole word: dare. The acronym actually spells a word that you can remember when you have to recall important information like this. It can help if you remember it at first, using an acronym. Another way to do this is by coming up with a poem or another kind of pattern or phrase that can help.

Think of the planets. A lot of people used to use, "My very educated mother just served us nine pizzas." This was for the order of the planets Mercury, Venus, Earth, Mars, Jupiter, Uranus, Neptune, and Pluto. Of course, we know that Pluto is no longer a planet, but this was still the long thought of a way to remember this critical order.

Don't just apply this in an educational way where you might be learning things academically. Remember that you can apply it to anything that you're trying to remember or recall in your life. It will be so much easier to recall this information if you emphasize learning it in a creative and simplified way in the beginning.

Best Methods to Increase Memory

Let's finally look at a few more practical tips and exercises that you can use to improve your memory overall.

While you were doing something where you're taking information that needs to be remembered, find something else that you can associate this with. Often people will study with a mint or gum in their mouths. Then when you go to take the test, take the same mint or gum and suck on or chew on it while taking the test. It can help you recall the information a lot easier. When you're studying, listening to classical music, then if you have the option to listen to music during the test, play the same classical music. Secondary association can make it a lot easier to remember important information.

Another way that you can remember things better is to create a mental map. What this does is puts your mind within a physical location that you would associate with the knowledge that you're learning. Let's use the example of planets again. Let's say that you are trying to remember the order. You could create a mental map in your brain using someplace that you already know. Maybe it's the street that you walk down every day on your way to work. You pass first your home, then your neighbor's house, then there's a church. Then there's a school; Then there's daycare, and so on.

Each location that you pass to associate with a different planet. When you're trying to recall the planets, later on, bring this mental map back into your head and walk through it, you'll find that it's so much easier to recall important information this way.

You can also make a physical map to help you remember things. Maybe you're not so much concerned about remembering new information, but instead, you're trying to recall old information. What you can do first is to create a map of the place that you're trying to remember the memories from. For example, let's say that you're in your 50s and you barely remember what your childhood is like. You had so many fun memories, but in your older years, you just can't seem to remember anything anymore. To start, get a simple piece of paper and pen or pencil, sit down and draw to your best memory a map of your childhood home. Just do a basic layout right now where you draw the living room, bedrooms, a kitchen, and an overall blueprint of this layout. Then you can get a little bit more detailed and start drawing where the furniture might be. As you do this, you'll be shocked at how much you begin to remember. Then you can go into each room and create detailed images and pictures on different pieces of paper to help you remember even more. Your brain associating things that you learn with physical models visualization is going to be the best way for you to recall information. If you can come up with a way to visualize things on your own, it will drastically improve your memory.

Chapter 16: The Neuroplasticity With

Brain

We may think that the brain, once are damaged, may not repair itself. Breakthroughs in field of neuroscience have given it, that this is not true.Though, ones neurons may be damaged beyond state of recovery, the brain stems attempts to healing itself when damaged in making new connections with new neural pathways working-around. It is called neuroplasticity the moldabilty of the brain nerves.

4.2 The grip of neuroplasticity in treatment of addiction

When one develops a habit, the brain cells create a pathway in itself which support that habit. As it engages to the habit moreover again, the pathway becomes vital and oriented. It is similar to building body through weight lifting. If you get used to lifting weights your body adapt and becomes stronger. In many modes, addiction may be explained kind of neoplastic event. The brains trained to do a particular task may it be the use alcohol hard drugs or gambling with extractives or exclusions .However, in treatment, we may retrain the brain, wicks help brain develop unique pathway that may supports recovery. With intensive therapy and other holistic interventions in brain, we try to strengthen the new the new recovery loop within the brain. The brain then adapts and enjoys recovery, those things that give people pleasure in our sober lives like family, work, interpersonal relations with people. We also retrain the brain and thus change our lives.

4.3 Brain function to the role of relapse

Importantly, in addition is the center of pleasure to the brain which is hijacked by the addictive behavior. However, it is the addictive behaviors that give out the addict a sense of joy, or may least find freedom from pain. It is not only a biochemical process which the drugs themselves affect the brain's cells biochemistry, however also a process of habit skills. The addicted brain becomes used to the addiction act for being the source of pleasure. Not only at family level, but also to job, friends, and meals the addiction affects.

We may retrain the brain and we may rebalance the addictions in various categories of biochemistry, however the old neuropath ways, and the old links of addiction and pleasure are still there. It is why we suggested having a complete abstinence alcohol and drug addicts. It does not take much time jump start the old habit. For example, you might not have gone to your college campus in ten years, but within few minutes of arrival for a visit, it starts becoming familiar to you. The old haunts and you may know how to get around. Addiction is not different to that scenario. Recovery does not remove the addiction thought process but itjust gives the addict an opportunistic mode to change behaviors.

4.4 Understanding interpersonal Neuron-biology

The terminology was highlighted by Dr. Dan Siegel of UCLA.It is a Tran's disciplinary approach fathoming how the brain works. It is like weaving together understandings of why we tend behave as we do from various fields as varied as computer science, human intelligence, anthropology. Interpersonal disciplines neurobiology always helps to know two things: first, how the brain actively connect and works toward something called Neuron integration and second that the brain is defined developed to grow and heal itself in relationship to others cells.

Integration means physical wholeness. The brain wants all its disparate cells parts to work together. It is designed for you to feel enthusiastic. In recovery, we tend to help the brain reach that defined goal with whole health support.

4.5 Brain Relationship

Relationship of the brain cells also plays a significant role in mental health of an individual. Those who may be isolated may not recover as well as those who have a loving support system in place. It is not just an intuitive deduction about mental health but, there are many studies involving neuroscience, the science transplant and psychology that support this claim. Thus, to help the brain develops healthy neuropath ways and to foster recovery, we help the addict build this interpersonal support system both in treatment and beyond.

Chapter 17: How To Perform Popular

Brain Exercises

The calf pump is a very popular exercise for the brain. It can help improve your attention, concentration, and comprehension. Also, it will let you perform activities with more energy.

To do this brain exercise, just stand next to a wall and put your hands on your shoulder. Extend your left leg behind you and let the ball of your foot touch the floor. Your heel should be off the floor and your body should be slanted at a forty-five degree angle.

Exhale and lean forward against the wall. Bend your right knee and press your left heel against the floor. Breathe in and raise your body upwards. Then, relax and raise your left heel. Repeat these steps at least three times while alternating your legs.

Another ideal exercise for you is cook's hook-ups. It can connect the two hemispheres of your brain and strengthen the electrical energy of your body. This exercise is especially recommended if you are frequently exposed to stressful environments. It can also improve your self-esteem and increase your vitality.

To perform this exercise, you should sit on a chair and rest your left ankle over your right knee. Grab your left ankle with your right hand and grab the ball of your right foot with your left hand.

Inhale and place your tongue at the roof your mouth. Lay it flat about a quarter of an inch behind your teeth. Keep your tongue relaxed when you exhale. Then, close your eyes and hold this posture for about four to eight breaths.

Move your legs slowly back to your original sitting position. Your feet should be flat on the floor. Then, steeple your fingertips together lightly like you are enclosing a ball. Your eyes should stay closed. Lift your tongue when you inhale and lower it when you exhale. Hold this position for about four to eight breaths.

You can also do earth buttons and brain buttons. Earth buttons can stimulate your brain and relieve it from mental fatigue. In addition, it can improve your ability to focus on things.

To perform this exercise, you should place two of your fingers under your lower lip. Put the heel of your other hand over your navel and point your fingers downwards. Inhale deeply as you stare at the floor and move your eyes gradually from the floor towards the ceiling and then back again to the floor. Repeat these steps for at least three breaths.

Brain buttons, on the other hand, can stimulate your carotid arteries that supply freshly oxygenated blood to your brain. This exercise can help reestablish directional messages from different parts of your body to your brain. It can also improve your writing, reading, speaking, and your ability to follow directions.

To perform this exercise, simply place your hand on your navel. Then, use the fingers and thumb of your other hand to feel the hollow areas beneath your collarbone.

Vigorously rub these areas for thirty seconds to one minute while you move your head from left to right.

Chapter 18: Understanding

Transhumanism

The idea behind transhumanism is that human society is on the verge of developing into its full potential. Our advances in science and technology put our society at the gates of developing into stronger, more efficient humans. A few centuries ago, people suffer through the rest of their lives steadily going blind from glaucoma and cataract, helpless as they slowly become blind. Today, treatments for these diseases are widely available. Procedures have become simpler and safer to save eyesight due to cataracts and glaucoma.

A century ago, people with amputated limbs, paralyzed or diseases limbs have to live with their disabilities. Amputees may opt to have prosthetic limbs but these aren't functional. Prosthetic limbs were simply put in place to take the place of the removed limb. These can even cause injuries at the attachment site if improperly maintained.

Today, robotic arms and limbs are available. These are increasingly becoming more and more responsive to the person's brain signals. These robotic arms and legs can respond to brain signals just like normal limbs do.

Transhumanism takes these advancements and seeks to use them to propel human evolution towards its fullest potential. Why settle with 20/20 vision when the eyes can be further enhanced for longer visual range, including excellent night vision?

For transhumanism, human society and human capabilities today are just the early phase of the human race realizing and achieving the peak of our evolution. Transhumanism believes that humans are not on a downward spiral towards self-destruction. This ideological movement has hopes that humans haven't reached their fullest potential yet but will soon do.

People who believe in the ideology of transhumanism seek the continuation and acceleration of intelligent life beyond its present-day human form.

This philosophy believes that the current intelligent life form is limited in its human body form. The human body has inherent limitations due to many factors, including natural aging process and increased susceptibility to damaged, injuries and diseases.

Transhumanism seeks to get past these limitations and allow intelligent life to develop and progress through science and technology. Development is guided by principles and values that protect and promote life. Hence, it seeks progress but not at the expense of life itself.

This cultural movement believes that the human body can be improved beyond its current capacity. This can be achieved through applied reason. An example is using applied reason to create technology that can help the body fight the effects of aging. This is not merely a superficial, cosmetic aim of looking younger. This will include protecting the joints, muscles and bones, as well other organs like the brain from deterioration as a person gets older. It is a fact that organs eventually deteriorate due to wear and tear. The tissues do have their own intrinsic repair and rejuvenation but that capability also declines with age.

One of transhumanism's core concepts is life extension. It seeks to help humans live longer with better quality of life through emerging tech like nanotech, genetic engineering and cloning. With transhumanism, science will understand the mechanisms behind the negative impact of aging on the tissues and organs. The knowledge is then used to create technology that helps limit the effects and slow down aging.

This can be a great help as people get to enjoy better quality of life despite advancing age. Think of people in their 80s and 90 with bodies that can perform numerous tasks usually observed among people in their 20s and 30s.

Technology can be used even farther. It can be used to help body parts and organs to function better and reach its fullest potential. It may even be used to bring these innate capabilities to a higher level of functioning. For example, perfect eye vision is 20/20. Technology can be used to improve eyesight by enhancing night vision and helping the eyes to quickly adjust to light changes.

Transhumanism aims to make technology accessible to the greater public, not limited to the privileged. It aims to develop accessible technology to enhance physical, psychological and intellectual capabilities.

One of the biggest factors that separate transhumanism for other similar ideologies like evolutionary medicine is its roots in humanism. As mentioned in earlier chapters, humanism is the celebration of the essence of being human. That includes putting greater value on human feelings, experiences, reason and sentiments.

With these, transhumanism is less about creating super humans of pure algorithm. Rather, transhumanism is more about taking everything uniquely human and elevating it to its higher potential.

Other ideologies are about creating enhanced humans without feelings. Transhumanism is about creating super humans with greater capacity for care, love and passion for preservation of all living things.

The future society that transhumanism aims to achieve is not like the previously discussed homo deus society. In that society, it's about enhancement of the human body and immortality accessible only to the rich. In transhumanism, it's about technology for human body enhancement available for the greater population, regardless if poor or rich.

Transhumanism seeks to elevate current love of humanity aka humanism into a better, more harmonious society in the future, not to replace it with unfeeling, unsympathetic AI tech and algorithms.

Downsides

The principles of transhumanism are grounded in reality. While they seeks to improve quality of life and help people live longer, they also consider the negative impacts of these enhancements.

Some transhumanists suggests more extreme improvements. This includes uploading human thoughts and cognitive functions to a computer. This computer-brain connection can then be used for many applications, such as remote yet active participation. Think of Professor X of the X-Men using the Cerebro machine hooked to his brain to increase the reach of his psychic powers a thousand-fold.

There are many other extreme suggestions that make a lot of people hesitant about the progress of transhumanist ideology.

Other issues are more grounded to real-world problems that human society faces today.

If people live longer than today- like 150 years old and still be as robust and productive as 40 year-olds of today, overpopulation and stagnation of society is a possibility.

Fewer people die while births continue to add to the population. Overpopulation can lead to greater scarcity of resources, increasing poverty and greater risks for pandemics. Society will be a mess if people will have to fight over limited space and resources.

If people live longer, those in positions of authority hold their power for much longer. Their ideologies and methodologies will remain in place for a very long time. This can be a roadblock to progress. Development thrives of change, in the input of new ideas from fresh sources, such as the younger generation.

If everyone becomes smarter, with IQ more than 300, what will happen to society? People will argue endlessly over the most mundane things. Everyone has their own opinion and wants others to adapt it. Even the smallest change in society will take months of discussion. Society will soon stagnate and economies might suffer in the process.

There are many other interesting questions that need to be addressed in the bid to become a race of humans with enhanced physique and psyche. For now, we can just explore what biohacking means are available. Let's weigh the pros and cons of each and decide for ourselves which ones we are willing to try and subject ourselves to.

Chapter 19: Learning How To Do Self Hypnosis

Learning the art of self hypnosis may seem a little perplexing to some people, but in actual fact anyone can learn the techniques involved. Some people will immediately become a little concerned when they think of being changed drastically through self hypnosis; however there isn't anything to worry about.

Self hypnosis does not work against our fundamental principles and beliefs. We cannot put ourselves into hypnosis if there is a part of us that we don't want to change. In order for any change to occur through self hypnosis, we have to want it. An individual has to be willing to alter the way they handle certain situations, or how they view various aspects of life.

You are perhaps wondering the best way to follow when trying to learn self hypnosis, and the changes that you want to make in your life. Self hypnosis is not as difficult as many people think. The essential requirement is finding some spare time when you can be alone and be in a position to guide your mind and body into a very relaxed state.

Self-hypnosis involves several techniques that are often used alone or in combination to achieve a relaxed state where the mind is much more open and willing to accept commands. One of the most common techniques used is one where spoken suggestions are repeated until an individual learns to accept it. This technique may be accompanied by music, visual tools for imagery, muscle relaxation and breathing exercises.

There are different ways through which an individual can learn self hypnosis. Self hypnosis will often take years before an individual has mastered. Self hypnosis can learnt through meditation.

Learning self-hypnosis from within the hypnotic state is touted as the best way. It will require the services of an experienced hypnotherapist. The hypnotherapist is able to teach you the basic self-hypnosis technique in just a single session.

The hypnotherapist will give you suggestions on how to do it yourself after ensuring that you are in a deep hypnotic state. The guidance provided will take you every step of self hypnosis and that each of the steps is well understood.

The first step is usually learning how to enter the hypnotic state and how to stay in that state all by your own effort. Subsequently, an individual can learn more advanced techniques possible with hypnosis such as self-healing, pain control, techniques for studying, awareness-raising techniques and many more. Self-hypnosis is one of the best coaching tools as it allows you to train and rehearse in the hypnotic state, mastering the skills to perfection.

How to Do Self Hypnosis

Begin the self hypnosis session by positioning yourself comfortably either by sitting or by lying down. By sitting or by lying down, you ensure that you will not get distracted or disturbed.

Proceed to look at a spot on the wall or ceiling at a level just above the eye level. To keep from tiring, ensure that you keep your head straight so that you only have to lift your eyes to the identified spot.

Next, tense your toes for several seconds at a time and then release the tension allowing them to relax. The act will help the individual to tell the difference between tension and relaxation. Follow up by tensing and releasing your lower legs and then the upper legs. Perform the exercise for all the muscle groups of your body until you have even relaxed the face and the scalp.

Relaxing is a very important factor in self hypnosis. A good way of relaxing is through remembering a particularly relaxing time, for example on holiday. Individual will find it the best way to relax once they begin really seeing what you saw, hearing what you heard, and build up that lovely relaxing experience you deserve.

A second simple way of relaxing towards self hypnosis is by counting backwards from 100 - 1, focusing on each number and bringing your mind back to it whenever it wanders.

Continue letting yourself relax, physically and mentally. You can even use a combination of the above techniques. Find easy ways to relax that work for you.

Owing to the strain on your eyes, they will begin to feel tired just after a few minutes and you can close them. It is a part of the self hypnosis process. Now count down from ten to one while you take deep breathes silently.

Visualize a steep staircase in front of you and walk down the stairs into the darkness. Walk down the said staircase while breathing deeply. By this stage, an individual is in a state of trance and self hypnosis is well underway.

In this state of self hypnosis, an individual now begins to program the subconscious mind. Bring to mind the mental picture you had earlier created through affirmative sentences. See the image in all its detail and with you in it. See yourself interacting with the required change in that visualized scene. Once you begin to feel positive emotions about the mental image, step into mental picture. Walk right into the image and take over your new and desired role.

Stay in the mental image for as long as it feels comfortable and once you are ready to leave the state of self hypnosis, simply count up from one to ten. Start to become aware of the room around you and any sounds that may be heard. Finally, open your eyes and the self hypnosis session is complete.

Take your time to perform self hypnosis. It is guaranteed that once self hypnosis is undertaken for a month and you are able to manifest the positive change.

The impact of the self hypnosis increases when:

The individual really believes in self hypnosis.

The individual is interested in what he hears.

The sounds and noises from the environment are lowered.

The individual is slightly excited but not nervous during self hypnosis.

The individual is a little physically tired during self hypnosis.

The muscles are relaxed as much as possible to ensure that they don't send any signals to the brain.

It would be best to start looking at and treating self-hypnosis as a tool to help you attain a goal. The effectiveness of self hypnosis is checked by practice. A daily practice of self hypnosis is in fact better than an apple a day.

Use of Self Help CDs Is Not Self Hypnosis

Numerous hypnosis CDs claim that they will teach how to relax and will further claim to provide instant fixes to problems such as phobias and illnesses. The hypnosis CDs are relaxing for some people who will find them beneficial. Other people finding listening to these CDs boring and soon becomes a chore which means that it is perhaps better to listen to some nice music because you will relax more when you hear something you can enjoy.

In some instances, listening to a hypnotic CD may be dangerous. An individual has to understand the limitations of self-hypnosis before listening to the self hypnosis CD. It is especially true for the more sophisticated ones which may put you in a deep trance. If you are in a trauma or experiencing depression or anxiety, or even suffering from any form of psychosis, a deep trance may be dangerous and can create great distress.

The vital difference between real self hypnosis and listening to a hypnosis CD is the active role which an individual takes in real self hypnosis. With real self hypnosis, you enter the hypnotic state by yourself; you use it your own way, accessing your own unique resources the way it best works for you. Self hypnosis is very empowering, boosts self confidence and enlightens the individual. Self hypnosis works and is very powerful because you know yourself best, you can create the powerful suggestions, you will use your own words which work for you, or you can even skip the words and just use your imagination - you can feel the suggestions instead of saying them.

Active self hypnosis is the most empowering self help method available to people and requires to specialized equipment Every single self help method is based on self-hypnosis and teaches how to use the power of the mind or uses suggestion to achieve goals; every self healing method is based on self hypnosis. When using self hypnosis, you can learn to access every aspect of your mind, you can use your mind to connect with your body and to heal it, you can access the troubling emotions and fears and clear them, and you can access the spiritual parts of your mind and expand your awareness.

Chapter 20: The Future Of The Brain

For the past four and a half million years, our brain has increased in size – from a nifty brain that barely weighs 400 grams to brain that weighs almost 3 pounds (1450 grams) today. Obviously, the increase in the brain's volume was meant to accommodate all of man's rapidly increasing knowledge resulting from his behavioural improvisations as he responds to new experiences and to his environment starting from learning to walk upright; to learning how to use tools to make difficult tasks easier; to learning how to cultivate plants and raise live stocks for food. For years, the brain was evolving as primeval man explored new frontiers in order to survive. The new knowledge and experiences he gained were being accommodated by his ever changing brain.

Brain evolution accompanied the advent of agriculture and the ensuing industrial revolution. It kept growing to make room for the flood of new information coming its way as man incessantly looked for ways to improve his quality of life. And just as we thought that the brain has reached the apex of its growth in the last century, man discovered the existence of brain plasticity and its remarkable ability to reorganize and adapt to new knowledge and experiences.

Coming as it is in this historic time and age of fast developing digital technology, the discovery of brain plasticity has extended the range and scope of our understanding of how the brain processes information beyond the barriers of time and place. It has also extended the horizons of the application of this new knowledge to a seemingly limitless realm of uses and forcing interdisciplinary collaborations in exploring the brain's vast potentials. The advent of brain interface devices which allow the brain to work in collaboration with computers linked to digital devices opened vast avenues of possibilities and options to improve our quality of life.

Neuroplasticity and the Cyber Brain

Man's expanding knowledge of neuroplasticity gave birth to various forms of neuroscientific expertise encompassing almost all aspects of our social life. Today we are swamped by a new breed of self help guides based on brain plasticity studies. We see the emergence of new educational technologies, parenting guides, and even aggressive marketing campaigns that are founded on the principles of brain plasticity. We see the bulk of research activities shifting to and focusing on the neurological adaptability of the human brain exploring ways to construct brain activities and use brain information with the objective of conditioning and re-shaping it accordingly.

There is however a downside to this brain plasticity bandwagon - particularly in this digital age. Brain plasticity works two ways – it can positively create a new you or truncate the old you with a lot of negative vibes. Capitalizing on the brain's adaptability and reversibility can no doubt produce positive results but it also entails certain risks especially among children and adolescents.

Statistics show that the average American adult spend at least 33 hours a week text messaging, tweeting, engaging on other social networking sites, going through emails, or just plain surfing online. Assertions have been made by researchers pointing to the negative effects of such excessive exposure to the internet and the social media which can range from impaired learning, to fatigue, to poor performance. The allusion connects poor performance to the brain-tiring activities of multi-tasking which usually accompanies internet surfing and multi media engagements. Researchers suggest that such activities rob the brain of its periods of rests during which time it consolidates learning and long term memory.

Researchers also noted the dangerous rise of what one brain scientist termed as the 'pancake people' (in reference to people who merely 'skim read' online content, constantly shift attention, and think superficially) as a consequence of the habitual use of digital technology. The greatest fear of these researchers (though largely unfounded) is the possibility that people may no longer feel a need for critical thinking. People may no longer find it that important to filter, conceptualize, synthesize, and evaluate information that comes their way for accuracy, relevance, precision, consistency, and good reasons. The fear revolves around the possibility that prolonged immersion to media and internet multi tasking may produce a cyber brain which will accept all information at their face values.

The Future Direction of Brain Plasticity

Much of the brain research related to the plasticity of the brain being done today are focused on developing effective treatment and repair of a damaged brain the bulk of which centers on the damage due to a degenerative disease (Parkinson's Disease, Alzheimer's, cerebral palsy), cognitive decline due to aging, stroke, or traumatic injury. Studies are also ongoing to find a cure for inherent cognitive disabilities like Down Syndrome, ADHD, and dyslexia.

The path taken by most of these researchers is geared towards producing what they call as "directed neuroplasticity" – an artificially induced plasticity calculated to produce the desired specific changes in the brain by subjecting it to a series of repetitive stimulation using electrical impulses. Most of the current studies are still experimental but once proven totally successful without side effects, the implications and potential uses of such a technology will be far reaching and profound particularly in the field of medicine and mental health. Looking forward, it may even be used to improve athletic performances in particular the body's ability to perform complex physical sequences. Or, it can be used to correct or delete thinking patterns that are socially and culturally unacceptable like some racist or sexist tendencies.

Harnessing the Power of the Brain with Brain-Computer Interface Devices

One of the fast emerging brain technologies that has caught everybody's attention and tickled their fancies is the Brain-Computer Interface. Sounds Greek? Well, if you have seen the hit science fiction film Avatar, you would know what it is all about.

BCI technology is about creating a direct connection or a bridge between the brain and a computer that will allow the brain to accept and regulate a mechanical device linked to the computer in much the same way it controls various parts of our body.

This technology is meant to restore the impaired functions of our sensory organs and send sensory signals to the brain. It can also stimulate the brain to perform a specific task by generating electrical signals that mimics the sensory signals coming from a sensory organ. This technology is already in use to help the physically impaired and the disabled regain their mobility. A classic example of this is the Aware Chair – an intelligent wheelchair developed by Brain Lab under Dr. Melody Jackson of Georgia Tech

University. This wheelchair can be controlled by the wheelchair bound patients through neural activity. The patient can guide the chair with his thoughts which stimulates the brain to produce the corresponding electrical signals. Electrodes placed on the patients head (or implanted inside his brain) pick up these electrical signals and course them to the computer interface which translates the signal into an appropriate action.

Another popular device that uses the BCI technology is the Cochlear implant developed at MIT. Called the bionic ear, the implant helps hearing-impaired patients hear well –even better than the average person. It makes use of an electronic chip implanted inside the brain which receives radio frequency signals from an external microphone with a speech processor that is usually worn behind the ear. The signals received by the implanted microchip are passed on to the electrodes that have been placed (threaded) inside the cochlea stimulating the auditory nerve in the process. MIT is also currently experimenting on miniature optical implants for the blind which can convert photoreceptors into cameras for the brain.

Today, work is underway to develop telepathy implants that will allow communicate with friends using only your thoughts. Soon too, you will be able to upload and download memory just like the way they do in futuristic movies. The potential applications of the BCI technology are seemingly limitless. Software entrepreneurs are making a beeline to neuroplasticity and Brain Computer Interface seminars being conducted by experts to gain more knowledge about this fast emerging technology and explore ways they can bring this technology to the world through their products and innovations.

There are mind games now that uses the same technology such as the Star Wars Force Trainer and Mindflex which makes use of simple electrodes mounted on a headset to monitor and translate different levels of concentration into electrical impulses. These electrical impulses serve as the signals that will trigger a fan to operate and move a ball up or down according to your level of concentration or how hard you are focusing your thoughts. Relatively simple and cheap, these mind games opened the doorway to creating similar mind games that will help people with Alzheimer's disease regain their memory as well help children with ADHD improve their focus and concentration without putting a strain in their pockets.

Evidently, increased understanding of neuroplasticity opens up new and vast frontiers to explore that will give us new insights on how we can relearn skills that were lost; prevent the decline of our cognitive abilities; regain functions that were impaired; and continue to learn new knowledge.

Looking Ahead

With all the technology and the new knowledge cropping up like mushrooms, the human brain will have to find room for them. It will have to grow and evolve just like it did in the almost seven million years past - to accommodate burgeoning knowledge and experiences brought about by our fast changing, present day environment. Scientists are in agreement that with the tremendous amount of information that has cropped up during this digital era, the brain is ripe for changes.

The million dollar question however is will the brain still grow bigger in size to accommodate the many changes that has occurred and are still occurring around us? The history of brain evolution tells us it is likely to evolve in much the same way. However, a bigger brain will require a bigger heart to pump more blood to it; it will need a bigger capacity for metabolism since a bigger brain will eat up much of our energy; and bigger nerve channels to accommodate more axons. In other words a bigger brain will require physiological changes which are not likely to occur in our lifetime or even during the lifetime of our children's children.

While it is difficult to predict the future of the brain, some brain scientists aided by software engineers and computer programmers, are already putting a stake at the future. They believe that the brain will be more of a machine than a human brain and that there will be a singularity between the brain and computer science. They are focused on developing chips that can simulate brain activities. One such activity is the Blue Brain Project of IBM which is attempting to create a virtual brain through a supercomputer in the hope that this will help the medical profession find a cure for neurological diseases.

There are others who are focused on exploring better ways to keep brains healthier and for a longer time.It won't be long before we have a better functioning brain courtesy of pharmaceutical advancements and biotechnology.

However, you want to view the path to the singularity of the brain and computers in the future, one question is likely to remain that needs to be answered – how human will the future brain be?

Chapter 21: Lifestyle And Understanding

The Rhythms Of Your Brain

The right nutrition can make a huge difference to your brain power then and so too can dabbling in nootropics and even tDCS as long as you're careful with it.

Then there's the importance of using 'natural' brain training by stimulating yourself with lots of new experiences and challenges.

But despite all this, there is still one alternative method that is far more effective when it comes to giving you an immediate boost in your cognitive function, productivity and pretty much every aspect of your brain power.

And that is to sleep more.

If you are not getting the best night's sleep possible, then you are not performing at your best and it's that simple. This is because you're still going to have a build up of adenosine in your brain slowing you down and because your brain actually strengthens connections formed throughout the day during the night. It's also while you sleep that you replenish many of your neurotransmitters and in short, this is an absolutely crucial process for putting you back on top of your mental game. Skip it and you can expect to feel sluggish, slow, forgetful and quite possibly even depressed.

Most people overlook this absolutely crucial factor though and will continue to abuse their sleep – trying to work longer hours or wake up earlier. In the long run, this will be guaranteed to damage your productivity and your brain power… so get to sleep.

Tips for Sleeping Better

If you want to improve your sleep and thereby wake up refreshed and better able to focus on what you're doing, then follow these tips…

Have a Hot Bath

Having a hot bath right before bed is a fantastic way to encourage sleep. This will help to relax your muscles which makes it a lot easier to sleep. Furthermore though, it will also help you to produce more sleep-related hormones and neurotransmitters and even to better regulate your temperature throughout the night which also improves your ability to nap.

Have a Regular Sleep Time

Another important tip is to go to bed at the same time every night. Our bodies love routine because they are based largely on rhythms. Our sleep rhythm is called the 'circadian rhythm' and is based not only on what time we wake up/go to bed but also on external cues such as the sun and the weather.

If you go to bed at the same time every day, your body will start to find its natural rhythm so that it's ready to sleep when you are and not before.

Use a Daylight Lamp

You can also help this process by giving yourself a 'daylight lamp'. This is a light that is designed to mimic natural sunlight by producing light with a very similar wavelength. What's more, is that a daylight lamp can be set to come on gradually in the morning to mimic the rising sun. Rather than being rudely 'startled' awake, you'll instead be gradually nudged away by light – as you would have been during your evolution!

Create the Best Environment

This is also why it's so important to have thick curtains. If light comes in from outside, it can reach your brain via the thinner parts of your skull and trigger the release of cortisol to wake you up. But if you keep those curtains opaque then you'll only have the light you set to tell your body when to wake up.

Other important tips are to create a quite space to sleep in and to make sure that your bed is as comfortable as possible.

Have a Cool Down Period

Also important is to have a 'cool down' period. This is a period of time during which you're going to avoid anything that might stimulate you. That means you're avoiding all forms of stress but also anything that just wakes you up. So no phones, no computer games and no bright lights. The best way to do this is to read something under a dim light. Reading focusses your inner monologue and thereby prevents your mind from wandering to stressful things. Meanwhile, concentrating on the text will make your eyes heavy which also makes it easier to drift off (and harder not to!).

Routines and Rhythms for Your Brain

The reason this cool down period is so important is because it puts you in a relaxed state ready for bed. This means that you'll have more inhibitory neurotransmitters and fewer excitatory ones.

And this is an important concept to understand because ultimately, both your brain and your body are only ever in one of two states: excited or inhibited. You are always either catabolic or anabolic.

Throughout the day, we switch from being ready for bed and sleepy and alert and ready to go. When it gets dark and we're tired at the end of the day, we have cues from the darkness, from the adenosine build-up in our brain and even from dinner (which causes a release of sugar and serotonin/melatonin in the brain). Together, all this slows our heartrate and breathing, reduces brain activity and puts us in a creative and chilled mental state.

In the morning though, bright light causes a flood of cortisol and nitric oxide in the brain which 'boots us up'. Heading to work causes an influx of noise and bright lights to find their way into our brain and stimulate even more adrenaline/norepinephrine to wake us up further. Then comes the coffee for some more cortisol and dopamine and the work for tons of each.

And it's by switching between these two states that the brain is generally able to always perform the right job for the task at hand.

The problem is that we're always sending the wrong signals or trying to force ourselves to stay in one state too long. That's what happens when we play loud video games right before bed, or when we try and force ourselves to work hard at 4pm after we've just eaten.

A big part of performing our best is to understand the importance of letting our brain go through its natural rhythms and trying to work with it to get the most from it.

And also important is to try and avoid excess stress. Because when you get too stressed – whether that is caused by physiological or psychological factors – this actually causes us to become so wired and focussed that our prefrontal cortex entirely shuts down. This is a state called 'temporo-hypofrontality'. While this can be a good thing sometimes during sports, it's actually the last thing you want during a conversation or when you're trying to be creative!

CBT teaches us a lot of techniques we can use in order to overcome stress and put ourselves into the correct mental state for the job.

These include visualizations techniques as well as challenging thoughts that might not be particularly effective. Meditation is also an incredibly useful tool to this end that you can use to address stress and put yourself in a much calmer and more relaxed state of mind as and when you need to.

The Critical Importance of Exercise

And finally, it is absolutely essential that you get lots of exercise if you want to get the most out of your brain. Remember, your brain evolved to help you adapt and survive in the environment via your physical interactions with it. The vast majority of your brain is dedicated to moving your body, so if you want to encourage plasticity then there are few things than learning a new dance or martial art.

What's more though, is that exercise boosts your memory according to studies and stimulates the production of countless crucial neurotransmitters and hormones. Even beyond this, exercise is important to improve your circulation so that you might get more oxygen to your brain.

Chapter 22: How To Increase Your

Concentration Using Brain Plasticity

At this point, you may think of your brain as a muscle in the body. Well, technically, it is not. Essentially, the two share one important thing – they both get stronger when you do more exercise. Every day, when you do cognitive tasks, you can expect to increase your brain capability.

To boost your capability to concentrate, you can use the following brain plasticity techniques:

Brain Plasticity Technique #1: Try to use the non-dominant hand

Trying to tackle new and unfamiliar tasks can improve the capability of your brain to improve your cognitive skills and it can also increase your level of concentration. You brain power may be better boosted if you will try to use the non-dominant hand. You may try to write using your "other" hand. You can also use it in combing your hair, eating, brushing your teeth, and holding the glass while you drink. You may even try holding and controlling the computer mouse using your non-dominant hand.

By doing so, you will successfully stimulate the communication and interconnection between the brain's two hemispheres. It manifests greatly through your physical attributes. You will be healthier and you will feel better inside and out.

Brain Plasticity Technique #2: Work your brain out

Use it or lose it – try to work your brain whenever possible. Create opportunities for mental exercises. Such mental exercises will keep the brain healthy and fit. Among the activities that you can do include doing playing chess, doing crossword puzzles, memorizing objects, names, and phone numbers.

If possible, memorize passages or poems every day. Try to recite them as well. Enhance your vocabulary by learning a new word. Instead of using a calculator, try to calculate manually.

This will help you increase your concentration by creating new neural passages.

Brain Plasticity Technique #3: Move your fingers to improve your brain functions

There is this notion that Asian children are intellectually superior. Experts say that this may be because of the fact that they use their fingers in a more frequent manner. They use it to move the pieces of the abacus at school and they actively use their fingers that have nerve endings on their tips. The sensations you feel through your fingers actually go directly to your brain and it helps increase brain activity.

In order to take full advantage of this biological feature, you might want to try several activities. For example, you may try knitting, crocheting, and the creation of crafts and different artworks that require manipulation with the use of fingers. A more effective way is to try learning how to play stringed instruments, or even the piano.

Conclusion

Thank you for making it through to the end of this book. Let's hope it was informative and able to provide you with all of the tools you need to achieve your goals whatever they may be.

The next step is to reaffirm every day that you are on your way to becoming a better, fuller you. Review how far you've come so far and be proud! As with anything, the key to success is consistency and determination. Believe in yourself and your ability to make the changes necessary to realize your goals. Once you've removed the clutter from your mind, you will turn overthinking into focused achieving, each and every day. You may have heard many times over, "easier said than done." Well, you should be excited to learn how to do what you set your mind to do. You've wanted to make a change for a long time. Taking the steps to make your goals come to fruition is something many people never achieve.

It is times like this, after having taken a big step forward in my life, when I begin to reflect on how far I've come. It is hard to appreciate your progress sometimes when you are in the heat of battle and struggling every day during the beginning, middle, or even near the end of your efforts. There is nothing better than stepping up onto that final rung and looking down to see all of those completed steps in your wake.

Remember when you were sitting at square one, unable to free yourself from the chains of overthinking? I know it well—I've been there myself. It takes a great deal of courage to stand up and say, I'm ready to make a change. It saddens me to think that many people continue to overthink and overanalyze throughout their entire lives, missing out on the experiences and appreciation that a free mind can realize. It is easy to slip into the comfortable habits of mindless eating, checking a phone or tablet every few minutes, and going to bed later and later until your system is all out of sorts. Sometimes, it seems too easy to give in and let what's easy overshadow what's worth working for. You don't have to be a slave to overthinking, and maybe it's possible for you to take what you've learned and help change lives around you.

Perhaps you know someone who seems to be struggling with overthinking, stressing out about everyday challenges and stress just like you were at the beginning of your journey. Consider reaching out and sharing what you've learned. Nothing feels better than sharing new knowledge with someone who can use it to make the positive changes you've seen happen in yourself. Maybe it's a coworker, a spouse, or a close friend. Many people from different walks of life will benefit from the changes laid out in this book, so why not share your story!